ATMOSPHERIC
ACOUSTIC
REMOTE SENSING

ATMOSPHERIC ACOUSTIC REMOTE SENSING

STUART BRADLEY

CRC Press
Taylor & Francis Group
Boca Raton London New York

CRC Press is an imprint of the
Taylor & Francis Group, an **informa** business

CRC Press
Taylor & Francis Group
6000 Broken Sound Parkway NW, Suite 300
Boca Raton, FL 33487-2742

First issued in paperback 2020

ISBN-13: 978-0-367-57755-1 (pbk)
ISBN-13: 978-0-8493-3588-4 (hbk)

Library of Congress Cataloging-in-Publication Data

Bradley, Stuart.
    Atmospheric acoustic remote sensing / author, Stuart Bradley.
        p. cm.
    Includes bibliographical references and index.
    ISBN 978-0-8493-3588-4 (hardback : alk. paper)
    1. Atmosphere--Remote sensing. 2. Echo sounding. I. Title.

QC871.B73 2006
551.5028'4--dc22
                                         2007034585

Visit the Taylor & Francis Web site at
http://www.taylorandfrancis.com

and the CRC Press Web site at
http://www.crcpress.com

Contents

Preface

In 2001 I was contacted by a consortium of research institutions and wind energy interests with a request to provide some background information on the operational characteristics of acoustic radars or SODARs. The consortium partners had set up and been funded for an European EU project to evaluate SODARs as a tool in monitoring wind flows at wind turbine sites. They felt reasonably confident in their knowledge of SODARs and had purchased some instruments, but wanted to be able to consult on any more complex issues which arose. Ultimately this developed into a relatively simple contract in which my colleague at the University of Salford, Sabine von Hünerbein, and I delivered an intensive two-day short course on SODARs to a small group of scientists and engineers at ECN headquarters in Amsterdam. There were two aspects of this short course which impressed themselves upon me. The first was the volume of information required to adequately cover the principles of operation and data interpretation for SODARs performing wind measurements in the atmospheric boundary layer. The second aspect was that intelligent and extremely technically capable people, already working in the area of wind measurement, did not adequately obtain enough information about remote sensing instruments from manufacturers' information manuals and data sheets.

The initial interaction with the 'WISE' EU consortium led to Sabine and me being responsible for overseeing the major calibration work-package in the project. The final report from the group working on that work-package was arguably the most comprehensive investigation of SODAR-mast calibrations. But, of necessity, that report was focused on wind energy applications and target goals for calibration accuracy. There still remained a need to make available a more general description of SODAR and other atmospheric acoustic remote sensing principles for a wider audience.

There is a huge body of literature available in journal papers which covers applications of acoustic remote sensing methodology in sensing atmospheric properties in the 1-km layer nearest the ground. But the body of literature describing design and operating principles is much more confined and also often rather specialized. Frequent requests from a range of scientists, engineers, local authorities, and other areas indicate that there is a demand for a more comprehensive collection of information on 'how things work'.

The difficulty in writing a book of this nature is to cover the principles of operation in detail sufficient enough that the reader is not left wondering about gaps in the descriptions, but at the same time trying to give a more intuitive feel for interactions between various atmospheric and instrumental components than might be found in a pedantically accurate textbook. Although the resulting book does contain considerable algebra, extensive use of diagrams makes for better readability and efforts have been made to avoid the more abstruse mathematical treatments.

SODARs and RASS instruments are endemic in monitoring atmospheric boundary layer wind systems, turbulent transports, and thermal properties. There is of course competition from other technologies such as mast-mounted cup anemometers and sonic anemometers, scintillometers, radar wind profilers, LIDARs, and radiometers. But it is still difficult to comprehensively replace, using these alternative technologies, the acoustic remote sensing capabilities of inexpensively providing wind, turbulence, and temperature profiles. As newer methods emerge, it is also very important to be able to competently compare output products.

For all these reasons, this book is aimed at providing a useful description of how atmospheric acoustic remote sensing systems work and giving the reader insights into their strengths and limitations.

Stuart Bradley
Auckland, New Zealand, 2007

Acknowledgments

Getting to the point of writing a book is a long road involving many years of personal interest in the research area targeted by atmospheric acoustic remote sensing. Clearly this is not really an individual effort at all, but rests on the enthusiasm and knowledge base of many colleagues as well as support and faith from family and friends.

The people at CRC Press, Taisuke Soda, Theresa Delforn, Jim McGovern and so many others, have been astonishingly patient with my many excuses and delays.

I am grateful to my many colleagues and friends in the "ISARS community" who meet every second year to share their latest insights at the International Symposia for Acoustic Remote Sensing. In particular the various organisers have added so much to this general field. Of these, my friend Bill Neff of NOAA has been a never-ending source of encouragement and inspiration (it is difficult to do something in this field which is entirely original and which hasn't already been visited by Bill in some way), Margo Kallistratova adds such useful insights while maintaining a twinkle in her eye, and Erich Mursch-Radlgruber is always so innovative with his latest hardware developments as well as knowing all the best wine cellars in Vienna! A seminal influence on this book has also been from the activities of the EU WISE consortium, and in particular Sabine von Hünerbein, Ioannis Antoniou, Detlef Kindler, Hans Jørgensen, and Manuel De Noord. A number of these people have gone on to form the nucleus of the EU UpWind remote sensing group.

All the manufacturers of SODARs and RASS have been happy to discuss aspects of their designs. In particular I would like to thank Hans-Jürgen Kirtzel and Gerhard Peters of Metek, and Ken Underwood of ASC for their friendship and many enjoyable analyses over a beer.

Many students and Post-doctoral Fellows at the University of Auckland and the University of Salford have worked happily with me in building my experience in acoustic remote sensing. I have really enjoyed the interaction with the group at University of Reading, and look forward to on-going interesting work with Janet Barlow on urban meteorology.

Sabine von Hünerbein of the University of Salford has worked extensively with me. Her pragmatic view of the science and questioning approach is refreshing and keeps me 'on my toes'. The two of us can seldom meet without getting into deep discussion about some aspect of SODAR or RASS operations, and I greatly value Sabine's input into aspects of this book.

Underpinning everything is Chrissie, my wife. She is so very patient in the panic times when deadlines loom, and endlessly encouraging in the quieter moments. When things become frustrating I know I can rely on talking with her and hearing her words of wisdom on how to tackle priorities. This book would not have been possible without her always being there for me.

Author

Stuart Bradley was born and educated in New Zealand, working as a researcher with the New Zealand Meteorological Service and with CSIRO in Australia before taking a position in the Physics Department at the University of Auckland. He designed one of the earliest phased-array mini acoustic radars and a number of more sophisticated instruments, as well as consulting for major manufacturers.

In 2000 Stuart accepted a Chair of Acoustics at the University of Salford in Manchester, UK, with a highly regarded acoustics research group. His work there included profiling of winds and temperature structure to better understand outdoor sound propagation; research relating to surface-atmosphere coupling in the Antarctic; and acoustic noise in cellular phones generated by micro-turbulence. He was a partner in the EU-funded 'WISE' project, which aimed to use acoustic remote sensing for evaluation of wind turbine performance. Together with Dr Sabine von Hünerbein at Salford and other collaborators in WISE, Professor Bradley produced a comprehensive analysis of the state of the art of acoustic wind profilers.

Stuart returned to the University of Auckland in mid-2004, where he is currently Head of the Physics Department. He holds a dual academic position in Salford where he is a Professor in the Acoustics Research Centre, engaged through the EU 'UpWind' project in design of new acoustic technologies for wind energy.

Symbol List

SYMBOL	DEFINITION
α	ratio of power spectrum variance to pulse spectral variance
α, α_e	absorption coefficient, excess absorption coefficient
β	scattering angle
β	fraction of pulse at each end used for shaping
χ_m, χ_h	similarity functions for momentum and heat
δ	fractional part of f/fs
ε	kinetic energy dissipation rate
ε	error vector
ε	ratio of molecular weight of water to molecular weight of dry air
ε_Θ	dissipation rate for heat energy
φ	phase
φ_x, φ_y	incremental phase shifts in x and y directions
Φ_n	power spectrum of refractive index fluctuations
γ	ratio of specific heats
Γ	gamma function
η	along-beam Doppler velocity component
ϕ	beam azimuth angle
$\phi_{R,T}$	azimuth angles from receiver, transmitter to point below scattering
Φ	latitude
Φ_V, Φ_T	kinetic energy, thermal spectral density
κ_m	von Karman constant
κ	spatial wavenumber
κ_D	Doppler-modified Bragg wavenumber
λ	wavelength
λ_a, λ_e	acoustic, electric wavelength
λ_D	wavelength of Doppler-shifted wave
λ_L	largest wavelength for a horn
Λ	drop size distribution parameter
μ	temperature ratio
θ	zenith angle
θ	beam tilt angle
$\Delta\theta$	width of main beam
θ_i, θ_r, θ_t	angles of incidence, reflection, and refraction
θ_R, θ_T	zenith angle for receiver, transmitter
θ_L	angular width of side lobe
Θ	potential temperature
ρ	air density
ρ	radial distance

ρ	spatial correlation coefficient
ρ_d, ρ_v	density of dry air, water vapor
ρ_R, ρ_T	distance of receiver, transmitter from point below scattering volume
σ_D	standard deviation in Doppler spectrum
σ_E	standard deviation of spectral noise
σ_f	standard deviation of frequency spectrum
σ_m	standard deviation of Gaussian pulse envelope
σ_m	standard deviation in slope
σ_w	standard deviation of vertical velocity
σ_R	rain scattering cross-section
σ_V	standard deviation of wind speed
σ_s	differential scattering cross-section
τ	pulse duration
υ	scaled height parameter
ω	angular frequency
Ω	angular velocity of rotation of Earth
$\xi,$	$\hat{f}_D / f_T$
ψ	wind direction
ζ	z/L
a	radius
a	amplitude
A	amplitude
A_e	antenna effective area for receiving
b	focal length
b	proportionality between absorption coefficient and f^2
B	matrix of beam directions
c, c_a	speed of sound
c'	fluctuating speed of sound
c_e	speed of EM radiation
c_p	specific heat
C	calibration matrix
C_n^2	refrative index structure function parameter
C_T^2, C_V^2	temperature and velocity structure function parameters
d	size of turbulent patch
d	speaker diameter or spacing
D	distance between transmitter and receiver
D	drop diameter
D	drift distance of RASS focus
D	distance from speakers to reflector
D	matrix of measured times
D	wind vector to radial component transformation
E	matrix of measurement errors
E	random noise amplitude spectrum
f	frequency
f_0	reference frequency

f_N	Nyquist frequency
f_{N_2}, f_{O_2}	absorption resonant frequencies for N_2 and O_2
f_s	sampling frequency
f_D	Doppler frequency
f_T	transmit frequency
F	matrix of frequency transformations
F	focal distance
g	acceleration due to gravity
G	antenna gain for transmitting
G_L	antenna gain for a side lobe
h	antenna height
h	water vapour molecular fraction
H	heat flux
H	Hanning filter function
i	spectral index
I	in-phase signal component
I	sound intensity
I_0	reference sound intensity
I_s	scattered intensity
j	$\sqrt{-1}$
J_0, J_1	Bessel functions
k	wavenumber
K	
K_h	thermal diffusivity
K_m	eddy viscosity
K_n	basis function for constrained fitting
l	layer spacing
l	mixing length
l_0	inner scale
L_E	Ekman length
L	Monin-Obukhov length
L_0	outer scale
L_p	sound pressure level (SPL)
L_I	sound intensity level
m	sample number
m	pulse modulation
m	slope of calibration
m	frequency step number
m, n	speaker numbers
M	number of samples
M	number of frequency steps
M	Fourier transform of pulse modulation function
M, N	numbers of speakers in each direction
M_{dryair}, M_{water}	molecular weights
M	average molecular weight

n	frequency bin index
n	refractive index
n_D	drop concentration
n_0	maximum drop concentration
n'	fluctuating refractive index
N	noise spectrum
N	matrix of along-beam velocities
N_0	reference noise spectrum value
N_m	number of speakers in row m
p	acoustic pressure
p	probability distribution function for wind speed
p_R, p_T	received and transmitted signals
p_{atm}	atmospheric pressure
p_{max}	maximum acoustic pressure
p_{rms}	root mean square acoustic pressure
p_0	reference acoustic pressure
P	power spectrum
P_A	power scattered by atmospheric turbulence
P_F	power reflected from fixed objects
P_r	Prandtl number
P_L	power received from a side lobe
P_P	power scattered by precipitation
P_N	noise power
P_R	received power
P_T	transmitted power
q	constant
Q	number of points in quadratic fit
Q	quality factor of BP filter
Q	quadrature-phase signal component
r	radial distance
r	relative humidity
r	regression coefficient
r_x	radius of a horn at distance x from the mouth
r_T, r_R	distance from transmitter, receiver, to scattering volume
R	radius of curvature
R	radial distance
R	rotation matrix
R	universal gas constant
R_d	specific gas constant for dry air
R_f	flux Richardson number
R_i	bulk Richardson number
s	signal
s	offset distance for speaker above dish
S	Fourier transform of scattering cross-section
SNR	signal-to-noise power ratio
t	time

t_s, t_r, t_s^*, t_r^* travel times in Doppler calculation

$t_{\text{downwind}}, t_{\text{upwind}}$ acoustic travel time downwind and upwind

T temperature

T beam tilt matrix

T matrix of unknown temperatures

T_D period of Doppler shifted wave

T_v virtual temperature

T, T^* period

T' fluctuating temperature

$\overline{T}$ mean temperature

$\underline{u}, \underline{v}, \underline{w}$ wind components

$\overline{u}, \overline{v}, \overline{w}$ mean wind components

u', v', w' fluctuating parts of wind components

$\hat{u}, \hat{v}, \hat{w}$ velocity component estimates

u_* friction velocity

v_m sampled signal voltages

V wind speed

V vector of wind components

V_s wind speed recorded by SODAR

V_c wind speed recorded by cup anemometer

$\overline{V}$ mean wind speed

$\overline{V}_\infty$ wind speed aloft

$\hat{V}$ wind speed estimate

V_0 reference wind speed

V_{rms} rms voltage

w water vapour mixing ratio

w_m amplitude weighting for speaker m

x, y, z spatial coordinates

X matrix of distances

$z_{\text{min}}, z_{\text{max}}$ minimum and maximum heights for signal reception

$\hat{z}$ estimated height

z_0 roughness length

Z length of reflector

Z_m mixing layer height

1 What Is Atmospheric Acoustic Remote Sensing?

A quick look at a bookshelf in a library on "remote sensing" will show many texts devoted to satellite sensing of clouds and the Earth's land and ocean surface. It is clear that satellite sensors are very remote, but does distance define "remote sensing"?

Audible sound is one of the human senses with which we are very familiar. Beyond that, we can also readily imagine the world of bats, cats, and dogs which use ultrasound and, with a little more difficulty, imagine sensing infrasound vibrations. In most cases, we use direct sound to communicate or detect something nearby rather than to sense any properties of the air around us. So what advantages does the use of sound have for investigating the atmosphere?

In this chapter, we first give some background information to answer these questions. Then the development of atmospheric acoustic remote sensing is briefly described, giving some idea of the status of this technology today, and future trends.

1.1 DIRECT MEASUREMENTS AND REMOTE MEASUREMENTS

A *direct* measurement is one in which the instrument is in close contact with, or close to, the object or region being measured. A *remote* measurement usually uses light, heat, or sound emitted from an object or region, and landing some distance away from the instrument. Remote measurements are characterized by the operator not being able to get close enough to do a "hands on" measurement. One feature of remote measurements is that there are often assumptions which need to be made about how the light, heat, or sound is emitted by an object or region, and how it travels across the intervening space to the instrument. An example is the RADAR (RAdio Detection And Ranging) in a speed camera. No actual contact is made with the car, and in fact the assumption is that the car reflected the RADAR signal rather than, for example, a bird or someone opening a window nearby. Assumptions are also made about what happens when it rains, for example.

All measurements contain errors. Errors can be *systematic*, which means that they are the same each time a measurement is made, or *random*, which means that they are different every time. There are good theoretical ways of handling random errors, as long as you know something about the probability of an error being of a certain size range. But systematic errors generally require good knowledge of the instrument and the ways in which the measurement was done. One of the biggest problems with remote sensing is in handling the errors and in estimating how "good" the resulting measurement numbers are.

Because of the need for assumptions in remote measurements, any new instrument needs to be compared to *ground truth*. This is the term used for direct measurements which can support the assumptions and calibrations applied to some remote measurement (the term comes from people going out into a field and doing direct measurements to compare with satellite remote-sensing measurements). At a new site, or periodically for older instruments, it is probably also desirable to occasionally do a ground truth measurement. For some meteorological instruments, such a comparison can be made using an instrumented mast. For other measurements, such as cloud thickness, ground truth comparisons can still be done (e.g., by an aircraft) but at much greater cost. Some of the most difficult ground truth comparisons are for planetary exploration.

Ground truth provides a *calibration*. Calibration gives the correspondence between the number provided by a remote measurement and the actual physical parameter it is trying to estimate. Usually it is assumed that the errors in the ground truth measurement method are much smaller than the errors in the remote measurement (otherwise the error analysis and working out what the measurement means get tricky).

There are a number of ways to take ground truth measurements. The first is a *field campaign*. This is where several scientists and technicians take a lot of equipment and set it up somewhere for a short but intense period of measurement. Often they go some place rather simple, such as the middle of flat land, an island or ship in the middle of the ocean, or an ice shelf at one of the poles. A lot of information is obtained from field campaigns, but they are expensive and only run for a short time.

1.2   HOW CAN MEASUREMENTS BE MADE REMOTELY?

To measure at a distance requires some form of information to travel from the *target* to the instrument. Usually this is light, heat, or sound but you could probably measure wind speed, for example, by measuring the shape of the splash mark made by a raindrop on blotting paper!

The first problem then is to be certain that the information came from the target region of interest. If you are doing measurements of sunspot activity, this is not a problem, but for many atmospheric remote-sensing instruments there is a mixture of information coming from the target and other sources. An acoustic sensor, for example, might have sound coming from the target, as well as sound from trucks, bees, trees rustling, and birds.

This means that some type of *matching* to the target is usually required. Simply making the remote-sensing instrument *directional* is often enough. For example, the speed-camera RADAR would be a very poor design if it was sensitive to things moving all around it, rather than focussing on the car on the roadway in front of the camera. So *beam-forming* is a part of the background needed to design a remote-sensing instrument. Secondly, most instruments try to match the energy emitted by the target to the type of energy to which the instrument is most sensitive. For example, measurements of ocean chlorophyll using a satellite will have a set of green color filters in the optics so that chlorophyll shows good contrast compared to seawater. This is *spectral-filtering* (the *spectrum* is the range of colors or wavelengths measured).

In addition to beam-forming and spectral-filtering, many remote-sensing instruments will also make use of time or rate-of-change information. The sound produced by a particular bird might be quite distinctive in duration compared with that from other birds in the neighborhood. Another example would be studying the activity of shrimp by listening to the underwater sound. If it is known that the shrimp are noisy when they are feeding at dawn, then all the acoustic data might be filtered to select only the hour before dawn and the hour after. This is using *temporal-gating*.

1.3   PASSIVE AND ACTIVE REMOTE SENSING

It is important to distinguish active sensors from passive sensors. A passive sensor measures the energy transmitted or reflected by some object when the source of energy is natural. An active sensor sends out a signal to illuminate the object and records or measures the response from the target or surface. Active sensors are largely independent of the environment around them and can operate 24 hours a day.

One of the best known active sensors is RADAR. RADAR is an acronym for RAdio Detection And Ranging. The scene to be "photographed" is illuminated with microwave radiation. Microwave images provide information about roughness and composition, which makes RADAR images more difficult to interpret than optical images. This book mostly concerns SODAR (SOund Detection And Ranging) which has many features in common with RADAR.

1.4   SOME HISTORY

John Tyndall first detected acoustic back-scatter from temperature and wind structures in the atmosphere in 1873 using a large fog horn. Little further work was done until 70 years later, when information about temperature inversions was required for early studies on microwave communications. Although an intriguing amount of atmospheric structure was observed in these 1946 measurements, and the term "SODAR" was coined, the interaction of sound with turbulence was not understood.

Theoretical work by Russian scientists Kallistratova and Tartarski in 1959 and 1960 established the principles of turbulent scattering of sound, but McAllister first showed that echoes could be reliably obtained from heights of several hundreds of meters and that the temporal structure of stable and unstable atmospheres could be analyzed through time–height plots of echo strength (McAllister, 1968). Much more theoretical and practical work followed, with a major advance being the first measurements of wind speeds based on the Doppler shift of echo signals (Beran et al., 1971). The significance of this advance at that time can be judged from the paper being published in *Nature*.

SODARs appeared to have reached the limit of their potential, until the design of compact high-frequency sounders together with digital output revitalized their use in meteorology (Moulsley and Cole, 1979). Design of both hardware and software rapidly improved so that today these instruments are portable, inexpensive, and very reliable if installed and operated correctly.

The use of a combination of an acoustic transmitter and RADAR, in what was called a RASS (Radio Acoustic Sounding System), first appeared in the early 1970s (North et al., 1973). These instruments use a transmitted acoustic wave as the target for a RADAR.

1.5   WHY USE ACOUSTICS?

Turbulence transports heat from higher-temperature regions to lower-temperature regions. This means that turbulent patches of air have a different temperature from their surroundings, and therefore a different density and refractive index n.

Refractive index changes cause scattering of light, microwaves, and sound. For example, turbulent scattering of starlight causes "twinkling." But the dependence of refractive index on temperature is very much weaker for electromagnetic radiation than it is for sound (Table 1.1). Note, however, that microwave radiation may also be refracted by humidity gradients.

Sound therefore reflects more strongly from turbulence. The generally lower cost of acoustic equipment makes sensing of atmospheric turbulence by sound even more attractive.

How large are the temperature variations in a turbulent patch? As discussed later, SODARs reflect efficiently from patches having diameters of half an acoustic wavelength, say 0.1 m. Even if a patch had the full background temperature variation (0.01 C/m) across it, this would mean a temperature variation of only 10^{-3} C and a refractive index variation of order one part in 10^6.

1.6   DIRECT SOUND PROPAGATION
FROM A SOURCE TO A RECEIVER

The SODAR and RASS instruments record energy scattered back from the atmosphere. An alternative remote-sensing methodology is to use direct sound transmission and to either monitor the intensity at a receiver, or monitor the travel time of the sound from the transmitter to the receiver. The former technique is really a part of the study of *outdoor sound propagation* and has not been very much used as a remote-sensing method, partly because of continuing disagreement between observations and models. New methods of signal coding and analysis may well make this a viable tool in the near future. The second method is *acoustic travel-time tomography* and has successfully been used to study turbulence and coherent structures moving across pastures.

The limitation of both these direct methods is that they essentially measure in the horizontal, and vertical profiles are not easily obtained. For example, an approximate guide to the height reached by sound energy which is directly measured in a horizontal propagation path is one-tenth of the horizontal range. This would mean a horizontal spacing of 1 km between the source and the receiver if refraction effects

---

TABLE 1.1

The sensitivity of electromagnetic and acoustic remote-sensing instruments to temperature variations

	Refractive index variation with temperature
RADAR	1.3×10^{-6} per °C
SODAR	1.7×10^{-3} per °C

were going to be used to sense atmospheric properties to a height of 100 m. Above these sorts of distances, the ground effects can be troublesome. In the case of travel-time tomography, refraction effects are undesirable, and this puts a limit on the likely horizontal range.

1.7 ACOUSTIC TARGETS

What are the reflectors when sound is sent vertically? Reflections of sound occur only if there is a change of density or air velocity, known as a change in acoustic refractive index. The air does generally change its density and wind speed slowly with height, but this type of continuous gradient is not sufficient to give measurable reflections. In practice, it is only when many reflections from small fluctuations add together that acoustic scattering can be measured. As we will see later, this is a calculable effect, and relates quite directly to the turbulent state of the atmosphere.

It should be noted that birds, insects, and precipitation within the acoustic beam can also act as reflectors. The occasional signal "spike" from a bird can readily be detected and removed by software filters, but precipitation is much more troublesome. Similarly, sound leaking from the instrument sideways and reflecting off a hard structure such as a building can swamp a valid reflection from turbulence. For these reasons, good understanding of the acoustic design of a SODAR's antenna is required. Antenna designs vary considerably, as seen from Figure 1.1 to 1.3.

1.8 CREATING OUR OWN TARGET

The principle of the RASS is to create the refractive index fluctuations deliberately and systematically by transmitting an acoustic beam vertically which is "tuned" so as to give strong reflections of the RADAR signal. This has the advantage that the

FIGURE 1.1 (See color insert following page 10) An experimental SODAR installed in 2004 at the British Antarctic Survey base at Halley in the Antarctic.

FIGURE 1.2 (See color insert following page 10) An AeroVironment (now Atmospheric Systems) SODAR conducting measurements of valley flows high in the Southern Alps of New Zealand.

FIGURE 1.3 (See color insert following page 10) Scintec's modern SFAS mini-SODAR antenna (left-hand photograph) and a Metek SODAR–RASS (right-hand photograph).

turbulent strength in the atmosphere is immaterial, and a signal is always received. The method has also been transferred to RADAR wind-profilers, with both the acoustic and RADAR beams transmitted at several angles to the vertical so as to sense various components of the wind.

1.9 MODERN ACOUSTIC REMOTE SENSING

SODARs and RASS now have a long pedigree as operational and research instruments. They are sufficiently compact to be carried in a small 4-wheel drive vehicle, and can generally be installed in a remote site within a few hours. Routine operation thereafter is really only a matter of periodically checking data quality in case there has been a change caused, for example, by a power failure, degradation of one or more acoustic transducers, or some physical misalignment occurring.

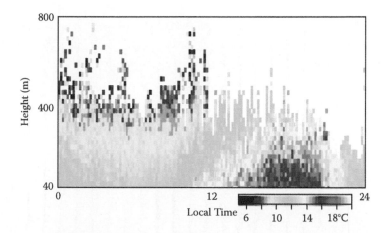

FIGURE 1.4 (See color insert following page 10) The atmospheric temperature recorded during one day by the Metek DSDPA**90.64** SODAR–RASS.

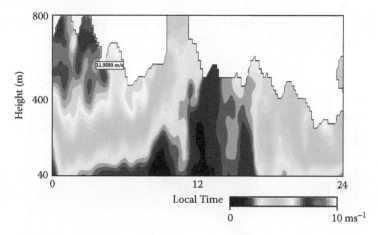

FIGURE 1.5 (See color insert following page 10) Wind speed profiles recorded during one day by the Metek DSDPA 90.64 SODAR.

The quality of data available is now very high. For example, Figures 1.4 and 1.5 show routine plots from the Metek SODAR–RASS of temperature and wind speed. Many other parameters can also be displayed.

Similarly, Figure 1.6 shows a plot of wind vectors recorded by an AcroViron-ment 4000 SODAR during a three-hour period.

Accuracy, with some care, can be better than 1% of wind speed, as shown in Figure 1.7.

1.10 APPLICATIONS

Applications are in all those areas where wind, turbulence, and temperature information is required. Figure 1.8 gives some indication.

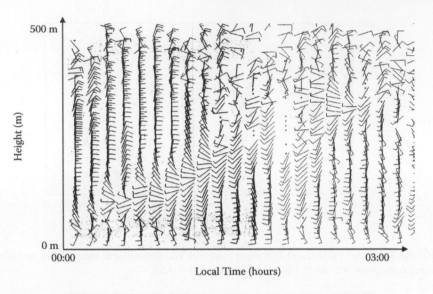

FIGURE 1.6 Wind vectors recorded by the AeroVironment 4000 during three hours. A strong upwardly propagating structure is observed.

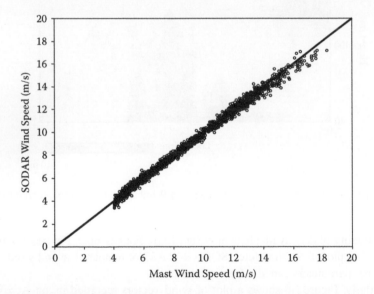

FIGURE 1.7 Correlation between SODAR measured wind speed and wind speed measured with cup anemometers on a mast.

1.11 WHERE TO FROM HERE?

Over the past few years, the challenges presented to atmospheric acoustic remote sensing include

1. greater accuracy of wind measurements for wind energy applications,

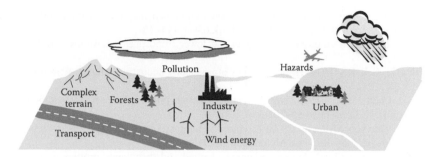

FIGURE 1.8 Some of the application areas for atmospheric acoustic remote sensing.

2. operation in urban areas, without disturbing the populace,
3. demand for better data availability,
4. easier installation and operation by non-experts,
5. reduction in the need for filters to exclude rain and spurious echoes from SODARs, and
6. desire for a more "turn-key" autonomous operation.

The route to meeting most of these challenges is in better design, and particularly acoustic design. For example, the new AQ500 SODAR has a parabolic dish design which is innovative, coupled with better acoustic shielding than has been seen in most other systems. The time is ripe for a quantum leap forward: this could be achieved through much more tightly specified acoustics, but could also come from new signal coding methodologies and from moving to a different acoustic frequency regime than used previously.

Given the often less-than-optimum use of these instruments, and the need for progressing toward new designs, the main scope of this book is to concentrate on describing the principles of design and operation of SODAR and RASS. This is the area in which it is more difficult to find research papers. There are many excellent reference sources for research applications, such as Asimakopoulos et al. (1996), Coulter and Kallistratova (2004), Kirtzel et al. (2001), Engelbart and Steinhagen (2001), Kallistratova and Coulter (2004), Neff and Coulter (1986), Peters and Fischer (2002), Peters and Kirtzel (1994), Seibert et al. (2000), and Singal (1997).

REFERENCES

Asimakopoulos DN, Helmis CG et al. (1996) Mini acoustic sounding — a powerful tool for ABL applications: recent advances and applications of acoustic mini-SODARs. Boundary Layer Meteorol 81(1): 49–61.

Beran DW, Little CG et al. (1971) Acoustic Doppler measurements of vertical velocities in the atmosphere. Nature 230: 160–162.

Coulter RL, Kallistratova MA (2004) Two decades of progress in SODAR techniques: a review of 11 ISARS proceedings. Meteorol Atmos Phys 85: 3–19.

Engelbart DAM, Steinhagen H (2001) Ground-based remote sensing of atmospheric parameters using integrated profiling stations. Phys Chem Earth Part B Hydrol Oceans Atmos 26(3): 219–223.

Kallistratova MA, Coulter RL (2004) Application of SODARs in the study and monitoring of the environment. Meteorol Atmos Phys 85: 21–37.

Kirtzel HJ, Voelz E et al. (2000) RASS — a new remote-sensing system for the surveillance of meteorological dispersion. Kerntechnik 65(4): 144–151.

McAllister LG (1968) Acoustic sounding of the lower troposphere. J Atmos Terr Phys 30. 1439–1440

Moulsley TJ, Cole RS (1979) High frequency atmospheric acoustic sounders. Atmos Environ 13: 347–350.

Neff WD, Coulter RL (1986) Acoustic remote sensing: probing the atmospheric boundary layer. D. H. Lenschow, pp. 201–239.

North EM, Peterson AM et al. (1973) A remote-sensing system for measuring low-level temperature profiles. Bull Am Meteor Soc 54: 912–919.

Peters G, Fischer B (2002) Parameterization of wind and turbulence profiles in the atmospheric boundary layer based on SODAR and sonic measurements. Meteorol Z 11(4): 255–266.

Peters G, Kirtzel HJ (1994) Complementary wind sensing techniques: SODAR and RASS. Ann Geophys 12. 506–517.

Seibert P, Beyrich F et al. (2000) Review and intercomparison of operational methods for the determination of the mixing height. Atmos Environ 34(7): 1001–1027.

Singal SP (1997) Acoustic remote-sensing applications. Springer-Verlag, Berlin, 405 pp.

FIGURE 1.1   An experimental SODAR installed in 2004 at the British Antarctic Survey base at Halley in the Antarctic.

FIGURE 1.2   An AeroVironment (now Atmospheric Systems) SODAR conducting measurements of valley flows high in the Southern Alps of New Zealand.

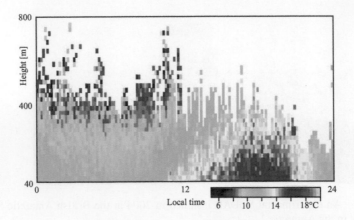

FIGURE 1.4   The atmospheric temperature recorded during one day by the Metek DSDPA SODAR–RASS.

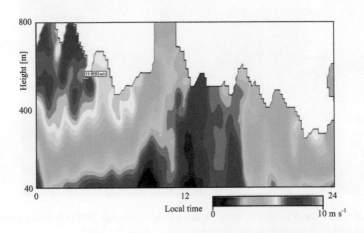

FIGURE 1.5   Wind speed profiles recorded during one day by the Metek DSDPA SODAR.

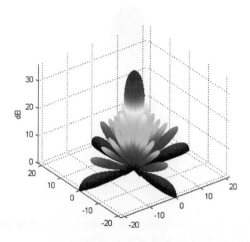

FIGURE 4.14   The beam pattern from an 8 × 8 square array without an applied phase gradient and with kd = 5.

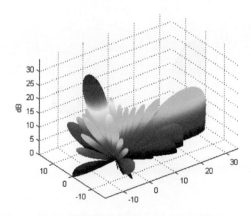

FIGURE 4.15   The beam pattern from an 8 × 8 square array with an applied phase increment of $\pi/2$ per speaker and with kd = 5.

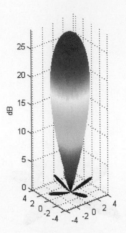

FIGURE 4.16   The beam pattern from an 8 × 8 square array without an applied phase gradient and with kd = 5 and a cosine-shaded speaker gain pattern.

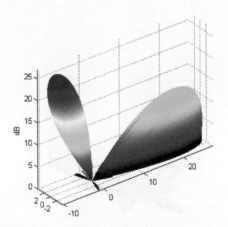

FIGURE 4.17   The beam pattern from an 8 × 8 square array with an applied phase increment of π/2 per speaker and with kd = 5 and a cosine-shaded speaker gain pattern.

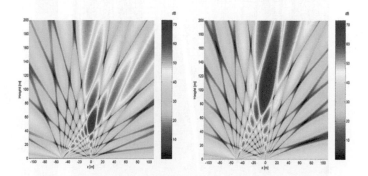

FIGURE 4.23   Unshaded bistatic system sensitivity for baseline D = 50 m, and with preset intersection height z0 = 50 m (left) and 100 m (right).

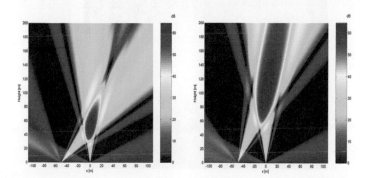

FIGURE 4.24   Shaded bistatic system sensitivity for baseline D = 50 m, and with preset intersection height z0 = 50 m (left) and 100 m (right).

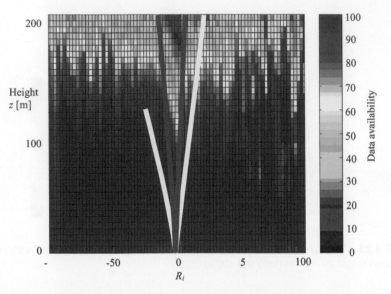

FIGURE 5.28 Percentage of relative data yield of Scintec SODAR receptions, plotted against height z of the SODAR range gates and against the Richardson number Ri based on meteorological mast measurements at 100 m. The solid yellow and blue lines are two contours of constant C_T^2 / Z^2.

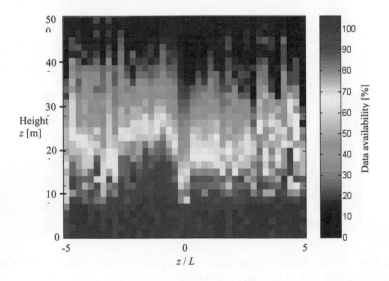

FIGURE 5.30 Data availability for the Metek SODAR based on Monin-Obhukov length L estimated from a sonic anemometer at 20 m height.

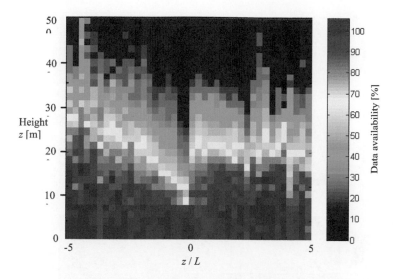

FIGURE 5.31   Data availability for the Metek SODAR based on Monin-Obhukov length L estimated from a sonic anemometer at 100 m height.

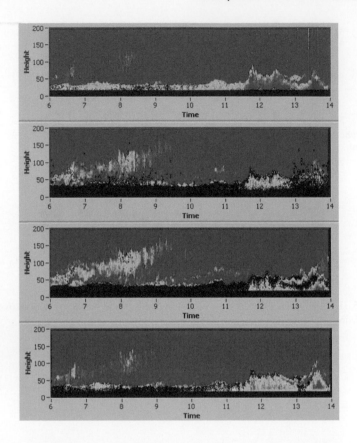

FIGURE 8.6 . Plots of time variations measured by the four SODARs of the C_T^2 field.

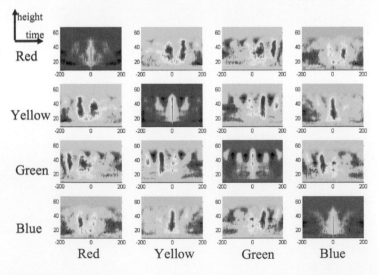

FIGURE 8.7 Matrix of covariances between C_T^2 values measured by each pair of SODARs at each height.

2 The Atmosphere Near the Ground

Atmospheric acoustic remote-sensing instruments are designed to give reliable measurements near the ground of atmospheric properties, such as wind speed, wind direction, turbulence intensity, and temperature. Estimates of these properties, in remote volumes up to several hundreds of meters above the ground, are obtained using a ground-based installation. Profiles of the atmosphere obtained in this way are then used for interpretation of atmospheric dynamics or transport mechanisms, and the results applied to the understanding of processes, such as wind power generation or urban pollution.

Effective design and use of acoustic remote-sensing instruments must therefore be coupled with some understanding of the lower atmosphere and the interrelationship between atmospheric properties. In this chapter, the structure of temperature, wind, and turbulence near the ground is discussed. Some general references covering this material are Blackadar (1998), Kaimal and Finnigan (1994), Panofsky and Dutton (1984), and Stull (1988).

2.1 TEMPERATURE PROFILES NEAR THE SURFACE

The atmospheric layer closest to the surface is strongly coupled to surface properties through friction. This friction-dominated *planetary boundary layer* is normally about 3 km deep, and vertical transport of heat, moisture, and momentum through the layer largely determines weather and climate. This is also the region most accessible to acoustic remote sensing.

Atmospheric pressure is due to the weight of air above, and so decreases with height. Near the surface, the air density ρ does not change significantly with height (a typical value is 1.2 kg m⁻³), the *hydrostatic* equation $\Delta p_{atm} = -\rho g \Delta z$ gives the pressure decrease Δp_{atm} for a height increase Δz, and g is the gravitational acceleration. If air rises or sinks, it will therefore expand or compress. Such pressure changes are accompanied by temperature changes, such as the heating that occurs in a bicycle pump when the air is compressed. However, because air is a poor heat conductor, vertically moving air does not exchange heat very effectively with the surrounding air. For a mass of air rising small distance dz, the change in potential energy g dz per unit mass of air is balanced by a change c_p dT in its internal heat energy, where c_p is the specific heat and dT is the change in temperature of the air. The result is the *adiabatic lapse rate*

$$-\frac{dT}{dz} = \frac{g}{c_p} = \frac{9.81\,\text{m s}^{-2}}{1005\,\text{J kg}^{-1}\,\text{K}^{-1}} = 9.8°\text{C cooling per km of altitude.}$$

Potential temperature, Θ, is a temperature measure with the natural 9.8°C per km removed, T is generally expressed in °C, and Θ is usually expressed in K. Near the surface, changes in the two temperature measures are related through

$$\Delta\Theta = \Delta T + (0.0098°\text{C/m})\Delta z \tag{2.1}$$

On average, cooling is less rapid than the adiabatic rate because of absorption by the air of heat radiated from the ground and because of mixing of air by turbulence. A typical lapse rate, used to define a *standard atmosphere* for computer models, is 6.5°C per km. So at any particular location and time, the *environmental lapse rate* will usually be different from the adiabatic lapse rate. If the environment cools more rapidly with height than 9.8°C per km, then rising air will be surrounded by cooler air and so will continue to rise: this is an *unstable* or *lapse* atmosphere and $d\Theta/dz < 0$. If the environment cools less rapidly with height than 9.8°C per km, then rising air will be surrounded by warmer air and so will sink: this is a *stable* or *inversion* atmosphere for which $d\Theta/dz > 0$. If the surface is heated by the sun, then air in contact will be hotter than the environment and will rise: this *convection* occurs during sunny days. When convective or wind-driven mixing of air is strong, the lapse rate will be close to adiabatic: this is the *neutral* atmosphere and $d\Theta/dz = 0$. Neutral and stable cases are shown in Figure 2.1.

A common occurrence is overnight cooling of the surface by radiating heat into a cold clear sky, and the cooling of the air closest to the surface through weak turbulent mixing. This creates a strongly stable layer of air in contact with the surface so that the environmental temperature initially increases with height. This region of increasing potential temperature with height is called a *temperature inversion*.

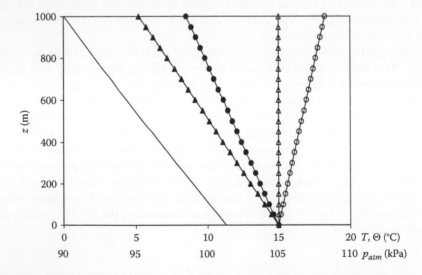

FIGURE 2.1   Height dependence of temperature, T (filled triangles for neutral or adiabatic case; filled circles for stable or average case), potential temperature, Θ (open triangles for neutral case; open circles for stable case), and pressure, p (solid line), in the lowest km.

At some height, the surface cooling effect will be insignificant and the temperature will again decrease with height. Inversions also occur at the top of fog where the droplets radiate heat in a similar manner to a solid surface, and also sometimes when one layer of air moves over another and their temperature structures are different. Inversions are important because pollutants, heat, and moisture become trapped in the underlying stable air.

Because the environmental lapse rate determines the vigor of vertical motion in the atmosphere, measurements of temperature profiles are very important in understanding and predicting atmospheric dynamics. Radio-Acoustic Sounding Systems (RASSs) are very useful as continuous measurement systems, whereas balloon soundings are generally used to obtain temperature profiles extending throughout the atmosphere.

2.2   WIND PROFILES NEAR THE SURFACE

Winds are slowed near the surface because of friction and obstacles, such as trees and hills. The action of different winds at two heights (wind shear) causes overturning which leads to smaller scale random motion or turbulence.

Usually, the wind velocity is visualized as consisting of components u, v, and w in the perpendicular x, y, and z directions, where x and y are horizontal (e.g., East and North) and z increases vertically. The total horizontal wind speed is $V = \sqrt{u^2 + v^2}$. The instantaneous components can be written as $u = \bar{u} + u'$, $v = \bar{v} + v'$, and $w = \bar{w} + w'$ where the mean value is shown with an overbar and the fluctuating turbulent value is shown with a prime (Fig. 2.2).

Even when there is no mean vertical velocity component, turbulent fluctuations w' will transfer heat, moisture, and momentum vertically. For example, the average rate at which the u momentum is transported vertically, per unit horizontal area, is $\overline{\rho u' w'}$. The quantity $\overline{\rho u' w'}$ is the average momentum flux of u momentum in the

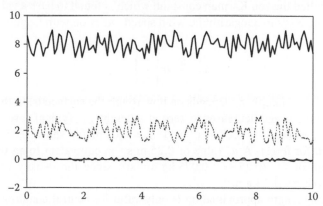

FIGURE 2.2   Typical time series of wind components u (dark line), v (dotted line), and w (thin line) showing mean and fluctuating parts.

z-direction, and its value determines what the turbulent connection is between the low-speed air near the surface and the higher-speed air aloft. The momentum flux can be measured directly by a sonic anemometer, which can measure simultaneously the 3-component wind fluctuations and form products such as $\overline{u'w'}$ and then average over a time interval.

In the lowest 10 m, a mixing length model describes this wind and turbulence interaction quite well. The small turbulent patches carry momentum from one level to another, but after moving a short vertical distance l (the mixing length) these patches break down. So the u' carried from level z to level $z + l$ accounts for the difference in the average horizontal wind speed at the two levels, or

$$\frac{d\bar{u}}{dz} = \frac{u'}{l}.$$

Also, if the turbulence is isotropic (the same in all directions), then $|w'| = |u'|$ and, allowing for the direction of transport,

$$\overline{u'w'} = -l^2 \left(\frac{d\bar{u}}{dz}\right)^2. \tag{2.2}$$

The simplest assumption is that the layer is a constant flux layer and therefore write

$$\overline{u'w'} = -u_*^2, \tag{2.3}$$

where the friction velocity u_* is a constant. The vertical distance in which overturning can occur is limited by how close the turbulent patch is to the ground, so it is assumed that

$$l = \kappa_m z, \tag{2.4}$$

where κ_m is called the von Karman constant, which is found to have a value of about 0.4. Integrating leads to a logarithmic wind speed variation with height:

$$\bar{u} = \frac{u_*}{\kappa_m} \ln\left(\frac{z}{z_0}\right). \tag{2.5}$$

The roughness length z_0 depends on how rough the surface is. In this approximation, the wind speed decreases to zero at a height z_0. Typical cases are shown in Figure 2.3, where $u_* = 0.25$ m s^{-1} and $z_0 = 0.01$ m for snow, 0.05 m for pasture, and 3 m for forest. A u_* value of 0.25 m s^{-1} is equivalent to an upward flux of momentum of $-1.2(0.25)^2 = -0.075$ kg m^{-1} s^{-2}, since the density of air at the ground is about $\rho = 1.2$ kg m^{-3}.

The mixing length approximation is only valid for neutral conditions and only in the lowest few tens of meters where the vertical flux of momentum is approximately constant. Above this constant flux layer, a useful approach is to assume that more momentum is transported vertically if the wind speed gradient is stronger. The

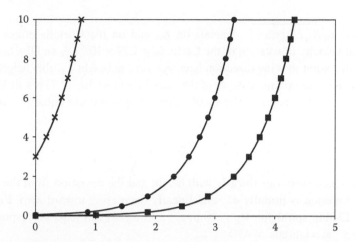

FIGURE 2.3 Height dependence of wind speed, u, from the mixing length theory, over trees (crosses), pasture (circles), and snow (squares) in the first 10 m above the surface.

assumption is therefore made that the momentum flux is proportional to the wind speed gradient or

$$\overline{u'w'} = -K_m \frac{d\overline{u}}{dz}, \qquad (2.6)$$

where the constant K_m is the *coefficient of eddy viscosity* and typically is 1 to 100 m^2 s^{-1}. Well above the frictional influence of the surface, the wind speed and direction are determined only by the pressure gradient (due to low-pressure or high-pressure systems) and the spinning of the earth (through the Coriolis effect). The latter effect means that the vertical flux of east-west momentum is fed into changes in the north-south wind component, and vice versa, leading to a twisting of the wind direction with height. This is the Ekman spiral, in which the wind $\overline{V}$ is small at the surface and its direction $\Delta\phi$ is 45° anticlockwise near the surface compared to $\overline{V}_\infty$ the wind aloft (or clockwise in the southern hemisphere). Equations for speed and direction are found to be

$$\overline{V} = \overline{V}_\infty \left(1 - 2e^{-\pi z/L_E} \cos\frac{\pi z}{L_E} + e^{-2\pi z/L_E} \right)^{1/2}, \qquad \Delta\phi = \tan^{-1}\left(\frac{e^{-\pi z/L_E} \sin\dfrac{\pi z}{L_E}}{1 - e^{-\pi z/L_E} \cos\dfrac{\pi z}{L_E}} \right),$$

$$(2.7)$$

where $L_E = \pi\sqrt{K_m/\Omega|\sin\Phi|}$ depends on K_m and on the Coriolis effect through the angular velocity of rotation of the Earth, $\Omega = 7.29 \times 10^{-5}$ s^{-1}, and the latitude Φ. Since the deviation in wind direction becomes zero at height L_E, this height can be considered as an approximate depth of the boundary layer. Eq. (2.7) is a little hard to interpret by visual inspection. For small heights z, approximate equations are

$$\bar{V} = \bar{V}_\infty \frac{\pi z}{L_E}, \qquad \Delta\phi = 45° \left(1 - \frac{2z}{L_E}\right) \tag{2.8}$$

so that the speed increases linearly with height and the deviation from the overlaying wind direction is initially 45° and linearly decreasing toward zero. Figure 2.4 shows an Ekman spiral for $\bar{V}_\infty = 10$ m s^{-1} and $L_E = 1$ km (e.g., corresponding to $K_m = 5$ m^2 s^{-1} at a latitude of 43°).

Although useful in models, both the mixing length and Ekman approximations are generally far too much of a simplification, and the wind structure near the surface needs to be measured or derived using additional information. This is one of the reasons that SODARs prove so useful.

2.3   RICHARDSON NUMBER

The momentum flux $\rho\overline{u'w'}$ is the rate of transfer of momentum per unit horizontal area. It therefore also represents a force per unit area or a stress. The product of force and velocity is a rate of doing work or a rate of change of energy. In the case of

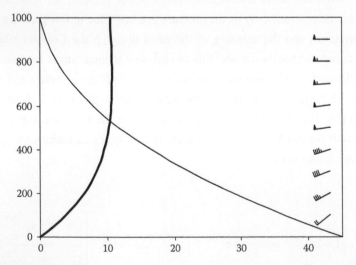

FIGURE 2.4   The wind speed in m s^{-1} (dark line) and wind direction in degrees (thin line) versus height in m for an Ekman spiral with a wind speed aloft of 10 m s^{-1}. The wind barbs on the right-hand side indicate direction from arrow barb to arrow point and speed by adding half barbs (1 m s^{-1}), full barbs (2 m s^{-1}), and filled triangles (5 m s^{-1}).

$$\overline{-u'w'}\,\frac{d\overline{u}}{dz} = K_m\left(\frac{d\overline{u}}{dz}\right)^2$$

the product represents the rate at which turbulent energy is transferred to the mean flow (per unit mass of air).

In a similar way to velocity fluctuations, the air temperature can be written as $T = \overline{T} + T'$ (in many of the definitions which follow, the fluctuation in potential temperature Θ' is commonly used instead of T': the two are essentially the same). An increase in temperature by amount T' means that the volume of air will be less dense and more buoyant (at constant pressure, density and temperature are inversely related). The force per unit volume on the air will be $\rho'g = T'(\overline{\rho}/\overline{T})g$. Again, the average rate of doing work by the buoyancy forces, per unit volume, will be the product of force and velocity, in this case $\overline{\rho w'T'}g/\overline{T}$. Per unit mass of air, this becomes $\overline{w'T'}g/\overline{T}$. The flux Richardson number is

$$R_f = -\frac{\overline{w'T'}g/\overline{T}}{K_m(d\overline{u}/dz)^2}. \tag{2.9}$$

When the numerator is positive, the temperature profile is unstable (warmer air is being carried upward by vertical velocity fluctuations), and R_f is negative. For stable temperature profiles, R_f is positive, and in this case the temperature stratification tends to reduce the turbulent fluctuations. When the flux Richardson number becomes greater than 1, the flow becomes dynamically stable and turbulence tends to decay.

The heat flux H can be written in terms of the temperature gradient through

$$H = c_p\rho\overline{w'T'} = -c_p\rho K_h\frac{d\overline{\Theta}}{dz} \tag{2.10}$$

and is used to define the (bulk) Richardson number

$$R_i = \frac{g}{\overline{T}}\frac{d\overline{\Theta}/dz}{(d\overline{u}/dz)^2} = P_rR_f. \tag{2.11}$$

2.4 THE PRANDTL NUMBER

$$P_r = \frac{K_m}{K_h} \tag{2.12}$$

has a value of about 0.7. Strongly stable air or low wind shear will therefore produce higher values of R_i and turbulence will be suppressed. In practice, it is found if a critical Richardson number of $R_i > 0.25$ is exceeded, then turbulence does not occur. Note that if R_i is negative, then the temperature profile is unstable. A single sonic anemometer can measure $\overline{u'w'}$ and $\overline{w'T'}$ directly but only infer $\mathrm{d}\overline{u}/\mathrm{d}z$. A SODAR can measure $\mathrm{d}\overline{u}/\mathrm{d}z$ directly and a RASS can directly obtain $\mathrm{d}\overline{\Theta}/\mathrm{d}z$, so the SODAR/RASS combination can obtain R_i profiles directly. The main reason for using the Ri, rather than Rf, is that the Richardson number includes terms which are much more easily directly measured.

2.5   THE STRUCTURE OF TURBULENCE

The *cascade theory* of turbulence assumes that the vertical gradient of the wind, or wind shear, arising from surface friction initially causes turbulent vortices of size L_0 (the *outer scale*). These are assumed to break down into successively smaller vortices (in a "cascade") until they become a small size l_0 (the *inner scale*) and then their energy is dissipated as heat. The essential idea is that turbulent kinetic energy (TKE) enters at large scales at a certain rate dependent on the generating mechanism, and is conserved as the turbulent scale sizes get smaller, and eventually dissipates at a rate ε per unit mass of air. The sizes of the turbulent patches are usually specified in terms of wavenumber $\kappa = 2\pi/\text{size}$. The idea of the cascade is that Φ_V, the kinetic energy (KE) per unit mass in a unit wavenumber interval, must be related to the rate, ε, at which KE per unit mass is dissipated as heat, and also related to wavenumber κ because there would be expected to be more smaller eddies in a volume than large ones. Assuming that ε appears to the power p and κ to the power q,

$$\Phi_V(\mathrm{J\,kg^{-1}\,m}) \propto \{\varepsilon(\mathrm{J\,kg^{-1}\,s^{-1}})\}^p \{\kappa(\mathrm{m^{-1}})\}^q.$$

Also $1\,\mathrm{J} = 1\,\mathrm{kg\,m^2\,s^{-2}}$, so equating mass, length, and time dimensions gives $\mathrm{m^3\,s^{-2}} = \mathrm{m^{2p-q}\,s^{-3p}}$ or

$$\Phi_V = 1.5\varepsilon^{2/3}\kappa^{-5/3}, \tag{2.13}$$

where 1.5 is an empirical constant. This is Kolmogorov's famous 5/3 law for the turbulent energy spectrum. In practice this means that, for turbulent patches with sizes in the range l_0 to L_0, the KE spectrum needs to be known at only one wavenumber interval in order to characterize the entire turbulent energy spectrum. However, the scale factor for the energy spectrum, $\varepsilon^{2/3}$, will vary at different sites and times depending on the rate at which turbulent energy is injected into the atmosphere at scales L_0 (typically 100 m), and lost as heat at scales l_0 (typically a few mm).

The turbulent energy dissipation rate ε is not easily measured directly. Dimensionally $\varepsilon^{2/3}$ is equivalent to $(\mathrm{m\,s^{-1}})^2\,\mathrm{m^{-2/3}}$ or a velocity squared divided by a length to the 2/3 power. A more easily measured quantity having this character is

$$C_V^2 = \frac{\overline{[V(x)-V(0)]^2}}{x^{2/3}} = \frac{1}{x^{2/3}}\left[\frac{1}{t_2-t_1}\int_{t_1}^{t_2}[V(x,t)-V(0,t)]^2\,\mathrm{d}t\right]. \tag{2.14}$$

Physically, C_V^2 is the *velocity structure function parameter* obtained by taking the time-averaged square of the difference in wind speed V at two points separated horizontally by distance x, divided by $x^{2/3}$. So in principle C_V^2 can be obtained directly from two sonic anemometers or possibly from a SODAR (this will be considered again later). The energy spectrum can now be written in the form

$$\Phi_V(\kappa) = 0.76 C_V^2 \kappa^{-5/3}. \tag{2.15}$$

At a particular site, a measurement of C_V^2 therefore characterizes the TKE, provided the scale is between the limits L_0 and l_0: stronger turbulence has a higher C_V^2.

Similarly, a reasonable assumption is that the temperature variance Φ_T, in a unit wavenumber interval, could depend on ε and κ, as well as on ε_Θ, the dissipation rate for heat energy:

$$\Phi_T[K^2\,m] \infty \{\varepsilon[m^2\,s^{-3}]\}^p \{\varepsilon_\Theta[K^2\,s^{-1}]\}^q \{\kappa[m^{-1}]\}^r.$$

This gives $K^2\,m = K^{2q}\,m^{2p-r}\,s^{-3p-q}$ from which $p = 1/3$, $q = 1$, and $r = 5/3$, so

$$\Phi_T = 0.106\varepsilon^{-1/3}\varepsilon_\Theta\kappa^{-5/3} \tag{2.16}$$

Again, it is more convenient to introduce a temperature structure function parameter

$$C_T^2 = \frac{\overline{[T(x) - T(0)]^2}}{x^{2/3}}, \tag{2.17}$$

which can actually be measured (by two Sonics, for example), so that

$$\Phi_T = 0.033 C_T^2 \kappa^{-5/3}. \tag{2.18}$$

The strength of mechanical turbulence and the magnitude of the temperature fluctuations are measured by C_V^2 and C_T^2.

The TKE dissipation rate is made up of two contributions: the rate of transfer of turbulent energy to the mean flow, $K_m(d\bar{u}/dz)^2$ and the rate of KE transferred into heat, $\overline{w'T'}g/\bar{T}$ or $-K_h(d\bar{\Theta}/dz)(g/\bar{T})$. So

$$\varepsilon = K_m\left(\frac{d\bar{u}}{dz}\right)^2 - K_h\frac{g}{\bar{T}}\frac{d\bar{\Theta}}{dz} = K_h\left(\frac{d\bar{u}}{dz}\right)^2(P_r - R_i) = K_h P_r\left(\frac{d\bar{u}}{dz}\right)^2(1 - R_f). \tag{2.19}$$

This again shows that $R_f < 1$ for turbulence to exist.

The dissipation rate for heat can be written as

$$\varepsilon_\Theta = K_h\left(\frac{d\bar{\Theta}}{dz}\right)^2. \tag{2.20}$$

Some relationships derived from these expressions, which will be useful later in the book, are

$$\varepsilon = \left(\frac{C_V^2}{1.97} \right)^{3/2}, \quad \varepsilon_\Theta = 0.22 C_T^2 C_V, \quad \frac{C_T^2}{C_V^2} = 1.63 \frac{R_i}{P_r - R_i} \frac{\overline{T}}{g} \frac{d\overline{\Theta}}{dz}. \tag{2.21}$$

2.6   MONIN-OBOUKHOV LENGTH

The cascade theory predicts that when $\varepsilon = 0$ there is no turbulent energy. From (2.19) this occurs when

$$K_m \left(\frac{d\overline{u}}{dz} \right)^2 = K_h \frac{g}{\overline{T}} \frac{d\overline{\Theta}}{dz}.$$

Rearranging this equation using (2.2)–(2.4) and (2.10) gives a height

$$L = -\frac{\rho c_p u_*^3 \overline{T}}{\kappa_m g H} \tag{2.22}$$

at which turbulence vanishes. This length is called the Monin-Oboukhov.

2.7   SIMILARITY RELATIONSHIPS

Very near the surface, turbulent processes are most likely shear-dominated and, for a surface layer having thermal stratification, further from the surface, turbulent processes are more likely to be buoyancy-dominated. The Monin-Oboukhov length is a useful estimator of the transition between these regimes since $L > 0$ for a stable atmosphere and $L < 0$ for an unstable atmosphere. The Monin-Obukhov similarity theory postulates that the shapes of the profiles of $\overline{u}$ and potential temperature Θ are functions only of the dimensionless buoyancy parameter

$$\zeta = \frac{z}{L}.$$

Therefore

$$\frac{\kappa_m z}{u_*} \frac{\partial \overline{u}}{\partial z} = \chi_m(\zeta), \quad \frac{\kappa_m z}{\Theta_*} \frac{\partial \Theta}{\partial z} = \chi_h(\zeta),$$

where χ_m and χ_h are empirically determined functions. One form of the *Businger-Dyer relations* is

$$\chi_h(\zeta) = \chi_m^2(\zeta) = (1 - 15\zeta)^{-1/2} \quad \text{for } \zeta < 0,$$

$$\chi_h(\zeta) = \chi_m(\zeta) = 1 + 5\zeta \quad \text{for } \zeta \geq 0. \tag{2.23}$$

Integration gives

$$\bar{u} = \frac{u^*}{\kappa_m}\left[\ln\left(\frac{z}{z_0}\right) + 5\frac{z}{L}\right] \quad \text{for } \frac{z}{L} \geq 0,$$

$$\bar{u} = \frac{u^*}{\kappa_m}\left\{\ln\left(\frac{z}{z_0}\right) - \ln\left[\left(\frac{1+v^2}{2}\right)\left(\frac{1+v}{2}\right)^2\right] + 2\tan^{-1}v - \frac{\pi}{2}\right\} \quad \text{for } \frac{z}{L} < 0, \qquad (2.24)$$

where

$$v = \left(1 - 15\frac{z}{L}\right)^{1/4}.$$

In the limit of a neutral atmosphere ($z/L = 0$), $\bar{u}$ is logarithmic with height. In the extremely unstable case,

$$\bar{u} \approx \frac{u^*}{\kappa_m}\left\{\ln\left(\frac{|L|}{2z_0}\right) + \frac{\pi}{2}\right\}$$

so log–linear profiles of the form

$$\bar{u} = a_1 + a_2 z + a_3 \ln(z)$$

apply in all cases (see Fig. 2.5). For potential temperature

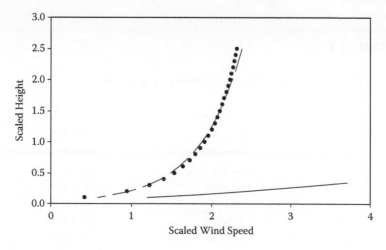

FIGURE 2.5 Log–linear variation of scaled wind speed $\kappa_m \bar{u} / u_*$ with scaled height z/L for $z_0 = 0.05$ m. Unstable atmosphere (dots), log–linear fit (dashed line), stable atmosphere (solid line).

$$\Theta = \Theta_0 + \frac{\Theta_*}{\kappa_m}\left[\ln\left(\frac{z}{z_0}\right) + 5\frac{z}{L}\right] \quad \text{for } \frac{z}{L} \geq 0,$$

$$\Theta = \Theta_0 + \frac{\Theta_*}{\kappa_m}\left(\ln\frac{z}{z_0} - 2\ln\frac{1+\upsilon^2}{2}\right) \quad \text{for } \frac{z}{L} < 0, \qquad (2.25)$$

where $\Theta_* = -H_0 / \rho c_p u_*$ and Θ_0 is effectively the potential temperature at height z_0. Again, both profiles are closely approximated by a log–linear profile.

2.8 PROFILES OF C_T^2 AND C_V^2

Measurements of profiles of both C_T^2 and C_V^2 are not common. Table 2.1 gives data from Moulsley et al. (1981) and these are plotted in Figure 2.6. As can be seen, both structure function parameters decrease with height, and $C_V^2 \approx 100 C_T^2$.

By including a model spectrum for the large boundary-layer scale eddies generated by atmospheric convection (w_*-scaling) alternative formulas for the structure function parameters are

$$C_V^2 = 3.9\frac{u_*^2}{z^{2/3}}\left[1+0.85\left(-\frac{z}{L}\right)^{2/3}\right], \quad C_T^2 = 4.9\frac{T_*^2}{z^{2/3}}\left[1+7.0\left(-\frac{z}{L}\right)^{2/3}\right]. \quad (2.26)$$

Wilson and Ostashev (2000)

TABLE 2.1

Profiles of C_T^2 and C_V^2 from Moulsley et al. (1981)

z (m)	C_V^2 (m$^{4/3}$ s^{-2})	C_T^2 (K^2 m$^{-2/3}$)
46	3.1×10^{-2}	6.31×10^{-4}
60	2.5×10^{-2}	6.0×10^{-4}
67	2.2×10^{-2}	5.0×10^{-4}
81	2.1×10^{-2}	4.0×10^{-4}
95	2.0×10^{-2}	3.0×10^{-4}
109	2.1×10^{-2}	2.21×10^{-4}
137	2.2×10^{-2}	1.51×10^{-4}
193	1.9×10^{-2}	8.0×10^{-5}
242	1.5×10^{-2}	5.0×10^{-5}

The SODAR frequency was 2048 Hz, beamwidth 9°, and T = 12°C.

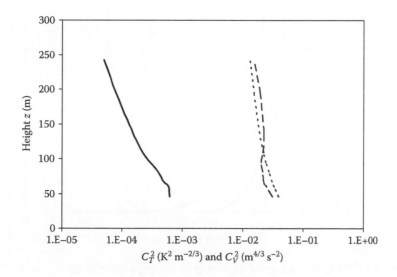

FIGURE 2.6   Profiles of measured structure function parameters C_T^2 (solid line) and C_V^2 (dashed line). A $z^{-2/3}$ line is shown for comparison (dotted).

2.9   PROBABILITY DISTRIBUTION OF WIND SPEEDS

When wind speeds are recorded over an extended period of time, the probability $p(V)dV$ of measuring a wind speed between V and $V + dV$ can be found. A model which matches experimental results quite well is the Weibull distribution

$$p(V) = \frac{q}{V_0}\left(\frac{V}{V_0}\right)^{q-1} \exp\left[-\left(\frac{V}{V_0}\right)^q\right].$$ (2.27)

For shape factor $q < 1$, the function decreases monotonically, $q = 1$ giving an exponential distribution with mean value V_0, and a maximum away from the origin appearing if $q > 1$. The mean wind speed is $\bar{V} = V_0\Gamma(1+1/q)$ and the variance is $\sigma_V^2 = [V_0\Gamma(1+2/q)]^2 - \bar{V}^2$. For example, choosing a shape factor $q = 2$ and scale factor $V_0 = 4$ m s^{-1} gives $\bar{V} = 3.5$ m s^{-1} and $\sigma_V = 1.85$ m s^{-1}, whereas a shape factor $q = 2$ and scale factor $V_0 = 6$ m s^{-1} give $\bar{V} = 5.3$ m s^{-1} and $\sigma_V = 2.8$ m s^{-1}. The corresponding distributions are shown in Figure 2.7.

2.10   SUMMARY

The key features of the atmosphere near the ground are

1. Temperature in a parcel of rising air cools at the adiabatic lapse rate of 9.8°C per km.
2. If the environmental air surrounding the air in a rising parcel is cooler, then the rising air is more buoyant, and the situation is unstable. If the

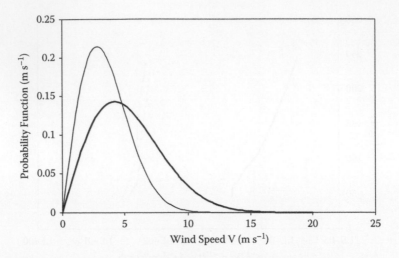

FIGURE 2.7 Weibull wind speed distributions having a scale factor $V_0 = 4$ m s^{-1} and shape factor $q = 2$ (thin line), and $V_0 = 4$ m s^{-1} and $q = 2$ (bold line).

environmental air is warmer, then the cooler air parcel will sink again, and the situation is stable. When the environment is adiabatic, then conditions are neutral and air can freely rise or sink.

3. Potential temperature gradient is a good way to visualize atmospheric stability.

4. Wind profiles near the surface are determined by vertical turbulence fluxes of heat and momentum. A model of a constant momentum flux layer is useful in the lowest 10 or 20 m. Above that level, some twisting of wind direction with height is often observed and the Ekman spiral theory gives some insight into this.

5. The Richardson number gives a measure of turbulent strength and for R_i above a critical value of 0.25 there is no turbulence.

6. Turbulent energy and turbulent temperature variance can be characterized by scale factors C_V^2 and C_T^2 and a wavenumber κ. The scale factors are related to fluxes and R_i.

7. The cascade theory leads to an estimate of the depth of the turbulent boundary layer called the Monin-Oboukhov length. Simple similarity relationships based on L allow average profiles of wind speed and temperature to be modeled.

8. The Weibull probability distribution is a good representation of wind speed statistics.

REFERENCES

Blackadar AK (1998) Turbulence and diffusion in the atmosphere. Springer-Verlag, New York.

Kaimal JC, Finnigan JJ (1994) Atmospheric boundary layer flows: their structure and measurement. New York Oxford University Press. 289 pp.

Moulsley TJ, Asimakopoulos DN, Cole RS, Crease BA, Caughey SJ. (1981) Measurement of boundary layer structure parameter profile by using acoustic sounding and comparison with direct measurements. Quart J Roy Meteor Soc 107. 203–230.

Panofsky HA, Dutton JA (1984) Atmospheric turbulence: models and methods for engineering applications.New York, Wiley, 397 pp.

Stull RB (1988) An introduction to boundary layer meteorology. Boston, Kluwer Academic Publishers, 666 pp.

Wilson DK, Ostashev VE (2000) A re-examination of acoustic scattering in the atmosphere using an improved model of the turbulence spectrum. Battlespace Atmospheric and Cloud Impacts on Military Operations (BACIMO), 25–27 April 2000, University Park Holiday Inn, Fort Collins, CO.

3 Sound in the Atmosphere

Acoustic remote-sensing tools use the interaction between sound and the atmosphere to yield information about the state of the atmospheric boundary layer. SODAR (SOund Detection And Ranging) and RASS (Radio Acoustic Sounding System) use vertical propagation of sound to give vertical profiles of important properties, whereas acoustic tomography uses horizontal propagation of sound to visualize the boundary layer structure in a horizontal plane. In Chapter 2, some of the fundamental properties of the turbulent boundary layer were discussed. In this chapter, the properties of sound are outlined. For a general coverage, see Salomons (2001). The primary interest here is what happens to the energy in a narrow acoustic beam directed into the atmosphere. In this case, the main effects are: spreading of the sound over a larger area as it gets further from the source; atmospheric absorption; sound propagation speed; bending of the beam due to refraction; scattering from turbulence; and Doppler shift of the received sound frequency. Discussion of diffraction over acoustic shielding and the reflection from hard surfaces will be left to a later chapter.

3.1 BASICS OF SOUND WAVES

When the flexible diaphragm of a speaker moves, it creates small pressure fluctuations traveling outward from the speaker. These pressure fluctuations are sound waves. The speed, c, at which these waves travel can be expected to depend on the mechanical properties p_{atm} (atmospheric pressure) and ρ (air density). A dimensional analysis, similar to those in Chapter 2, shows that

$$c \propto \sqrt{\frac{p_{atm}}{\rho}} \tag{3.1}$$

and, as already noted, the temperature and density are inversely related to each other at constant pressure through the gas equation

$$p_{atm} = \rho R_d T,$$

where $R_d = 287$ J kg^{-1} K^{-1}. This means that

$$c \propto \sqrt{T}. \tag{3.2}$$

Allowing for T being the temperature in K, and that the speed of sound at 0°C is 332 m s^{-1},

$$c(T) = 332(1 + 0.00166\Delta T)\,\text{m s}^{-1}, \tag{3.3}$$

where ΔT is the temperature in °C. For air containing water vapor, the air density is the sum of the dry air density, ρ_d, and the water vapor density, ρ_v, or

$$\rho = \rho_d + \rho_v = \frac{p_{atm} - p_v}{R_d T} + \frac{p_v}{(R_d / \varepsilon) T},$$

where $\varepsilon = 0.622$ is the ratio of the molecular weight of water to molecular weight of air, and individual gas equations have been used for dry air and for water vapor. A simpler expression is obtained in terms of the water vapor mixing ratio, $w = \varepsilon p_v / (p_{atm} - p_v)$, which is the mass of water vapor divided by the mass of dry air per unit volume. Rearranging gives

$$\frac{p_{atm}}{\rho} = R_d \left(\frac{1 + w/\varepsilon}{1 + w} \right) T = R_d T_v,$$

where T_v, the virtual temperature, allows for the slight decrease in density of moist air. More precisely, the adiabatic sound speed is

$$c = \sqrt{\frac{\gamma R T}{M}},$$

where $R = 8.31$ J mol^{-1} K^{-1} is a universal gas constant, γ is the *ratio of specific heats* for the gas, and $\bar{M}$ is the average molecular weight. This sound speed does not allow for the effect of air motion (i.e., wind) in changing the speed along the direction of propagation. When a fraction $h = p_v/p_{atm}$ of the molecules is water vapor, both γ and $\bar{M}$ depend on h via

$$\gamma = \frac{7 + h}{5 + h}, \qquad \bar{M} = M_{dry\ air}(1 - h) + h M_{water}.$$

These expressions interpolate between $\gamma_{dry\ air} = 7/5$ and $\gamma_{water} = 8/6$, and also between the two molecular weights. After a little algebra, and allowing for the fact that $h \ll 1$,

$$c = \sqrt{\frac{\gamma_{dry\ air} R T}{M_{dry\ air}} \left[1 + \left(1 - \varepsilon - \frac{2}{35} \right) \frac{e}{p} \right]} \approx \sqrt{\gamma_{dry\ air} R_d T_v}.$$

If the sound pressure disturbance is traveling in the $+z$-direction, then the wave can be described by

$$p = p_{max} \cos(\omega t - kz + \varphi) = \sqrt{2} p_{rms} \cos(\omega t - kz + \varphi), \qquad (3.4)$$

where the amplitude p_{max} of the acoustic pressure variation is much less than the typical atmospheric pressure of 100 kPa. It is also useful to write this expression as a complex exponential

$$p = p_{max}\, e^{j(\omega t - kz + \varphi)} \qquad (3.5)$$

where $j = \sqrt{-1}$ and the physical sound pressure is the real part.

The angular frequency ω is related to the sound frequency f and the period T of oscillation through

$$\omega = 2\pi f = \frac{2\pi}{T} \tag{3.6}$$

and the wavenumber k is related to wavelength λ and sound speed through

$$k = \frac{2\pi}{\lambda} = \frac{\omega}{c}. \tag{3.7}$$

The phase angle φ allows for the pressure not necessarily being a maximum when $t = 0$ and $x = 0$. Typically a SODAR frequency is $f = 3$ kHz, and for $\Delta T = 15°C$ the sound speed is $c \approx 340$ m s^{-1}, wavelength $\lambda = 0.11$ m, $k = 55$ m^{-1}, $\omega = 18850$ s^{-1}, and period $T = 0.33$ ms. Figure 3.1 gives an illustration of sound wave parameters.

The root-mean-square (RMS) pressure value, P_{rms}, is a useful measure of the size of disturbance for any periodic wave shape, and is defined by averaging the square of the pressure variation over one period, and then taking the square root

$$P_{rms} = \sqrt{\frac{1}{T} \int_0^T p^2 \, dt}. \tag{3.8}$$

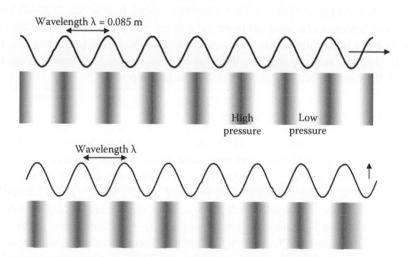

FIGURE 3.1 An acoustic pressure wave of frequency 4 kHz and pressure amplitude 0.2 Pa traveling from left to right with speed of sound 340 m s^{-1}. The upper plot shows pressure versus distance at time $t = 0$ and below that a visualization of the compressions and rarefactions in the air along the longitudinal wave. The lower plot shows the pressure variations a quarter period or 62.5 µs later, during which time the wave has traveled a distance $cT / 4 = \lambda / 4 = 21.25$ cm.

Because of the wide dynamic response of the human ear, it is common to use a logarithmic scale for sound intensity. The *sound pressure level* measured in dB (decibels) is

$$L_p = 10 \log_{10} \left(\frac{p_{\text{rms}}^2}{p_0^2} \right),$$ (3.9)

where the reference pressure $p_0 = 20$ μPa is the very small rms pressure fluctuation which is at the threshold of hearing. Note that sound intensity is proportional to the square of the pressure amplitude, which is why pressures are squared in (3.9). At the other extreme of intensity is the threshold of pain, for which $L_p = 120$ dB (or $p_{\text{rms}} = 20$ Pa). In practice, the human ear has some frequency sensitivity and a modified scale can be used with "a weighted response" and measured in dBA to allow for this. But in the case of SODAR, RASS, and tomography, the interest is generally in the response of transducers and so L_p is used, or alternatively a logarithmic intensity level

$$L_I = 10 \log_{10} \left(\frac{I}{I_0} \right)$$ (3.10)

also measured in dB, where I is the sound intensity in W m^{-2} and the reference intensity corresponding to the threshold of hearing is $I_0 = 10^{-12}$ W m^{-2}. For example, if a SODAR is transmitting 1 W of acoustic power, then at 1 m from the source, the 1 W is spread over an area of 4π m^2 giving an average intensity round the entire SODAR of $1/4\pi$ W m^{-2}. The intensity level would be $L_I = 10 \log_{10}((1/4\pi)/10^{-12}) = 109$ dB. This is only meaningful if the sound is omnidirectional: in practice, SODAR transducers and antennas are designed to be very directional, and so the intensity level could be much higher directly in the acoustic beam. Also it is important to note that *acoustic power* is referred to, since the total electrical power delivered to a speaker is generally much higher than the transmitted acoustic power.

3.2   FREQUENCY SPECTRA

Background acoustic noise, the received echo signals, and even the transmitted signal are not composed of single-frequency sinusoidal waves. It is therefore useful to record and plot *frequency spectra* which show how much acoustic power there is per unit frequency interval. Since the phase of the received sound is usually not of interest (an exception is acoustic travel-time tomography), *power spectra* are usually recorded.

Suppose that an acoustic pressure $p_0 \cos(2\pi f_0 t)$ is recorded in a narrow frequency band Δf centered on frequency f_0, together with other values at other frequencies. If we multiply the entire input signal by $\cos(2\pi f_0 t)$ and integrate over a long time then the result for the band around f_0 is $p_0 \Delta t / 2$. For any other frequency f_1, the gradual phase shift between $\cos(2\pi f_0 t)$ and $\cos(2\pi f_1 t)$ means that their product averages to zero. In this way, each individual spectral density component can be recovered from any general signal. The method is generalized using complex exponential notation, and taking

$$P(f) = \int_{-\infty}^{\infty} p(t) e^{-j2\pi ft} \, dt, \tag{3.11}$$

which is known as the *Fourier transform* of a signal $p(t)$, and the *inverse Fourier transform* is

$$p(t) = \int_{-\infty}^{\infty} P(f) e^{j2\pi ft} \, df. \tag{3.12}$$

In practice, signals are invariably sampled at discrete times $m\Delta t$ ($m = 0, 1, 2, \ldots,$ $M - 1$), so

$$P(f) = \int_{-\infty}^{\infty} p_m \, e^{-j2\pi fm\Delta t} \, dt \approx \sum_{m=0}^{M-1} p_m \, e^{-j2\pi fm\Delta t} \Delta t.$$

For symmetry in the inverse transform, the power spectrum is also estimated at discrete frequencies $m\Delta f$ ($m = 0, 1, 2, \ldots, M - 1$), so (omitting the Δt)

$$P_n = \sum_{m=0}^{M-1} p_m \, e^{-j2\pi mn\Delta f\Delta t}, \quad n = 0, 1, \ldots, M - 1. \tag{3.13}$$

Within the total sampling time of $M\Delta t$, the lowest frequency having a complete cycle is $\Delta f = 1/(M\Delta t)$. The highest frequency in the power spectrum is therefore $M\Delta f = 1/\Delta t$. However, at each frequency interval the signal has both an amplitude and a phase (with respect to $t = 0$), so spectral densities at frequencies from $1/(2\Delta t)$ to $1/\Delta t$ are really just further information about the signal components in frequency intervals from 0 to $1/(2\Delta t)$. For this reason, the highest frequency recorded, called the *Nyquist frequency*, is $f_N = 1/(2\Delta t)$. The sampling frequency is $f_s = 2f_N$, or in other words the signal is sampled at twice the highest frequency for which a spectral estimate is obtained.

What if the original signal contained components at higher frequencies than f_N? These are frequencies for which $n = M+q$ in (3.13) where q lies between $-M/2$ and $M/2$. From (3.13)

$$P_n = \sum_{m=0}^{M-1} p_m\, e^{-j2\pi mn/M}$$

$$= \sum_{m=0}^{M-1} p_m\, e^{-j2\pi m(M+q)/M}$$

$$= \sum_{m=0}^{M-1} p_m\, e^{-j2\pi mq/M}\, e^{-j2\pi m}$$

$$= \sum_{m=0}^{M-1} p_m\, e^{-j2\pi mq/M}\, (\cos 2\pi m - j\sin 2\pi m)$$

$$= \sum_{m=0}^{M-1} p_m\, e^{-j2\pi mq/M}$$

$$= P_q.$$

This means that any signal components having frequencies above f_N appear at lower frequency positions within the spectrum. This is called aliasing. Aliased components add to the components which are really at a lower frequency, and this can cause a very distorted impression of the true spectrum. For this reason, low-pass anti-aliasing filters should be used to remove all signal components above the Nyquist frequency, prior to digitizing the signal. An example of aliasing is given in Figure 3.2 where $f_N = 2000$ Hz. Note that when a signal component is at $f_N + 500$ Hz, it adds to any other components at $f_N - 500$ Hz. In this MATLAB®-generated plot, the spectral density scaling for the FFT routine is $N/2$.

There is a very efficient method, called the fast Fourier transform (FFT), for doing the sums required to perform the Fourier transform.

3.3   BACKGROUND AND SYSTEM NOISE

An acoustic remote-sensing system must detect signals in the presence of background and system noise. Random noise sources include electronic noise from the instrument's circuits, and acoustic noise from the environment. In addition, unwanted reflections from nearby buildings or trees ("fixed echoes") can obscure a valid signal, but these are not random noise.

Electronic noise comes from the noise in the preamplifier, from resistors near the front end of the instrument's amplifier chain, and from microphone self-noise. It is most important that these noise sources are minimized, since noise voltages from this point receive the greatest amplification. A good operational amplifier can have typically 1 nV Hz$^{-1/2}$ referred to its input. This means that if the bandwidth is 100 Hz, then the equivalent rms noise voltage at the input of the operational amplifier is 10 nV. Input resistors, and the resistance in the speaker/microphone, also con-

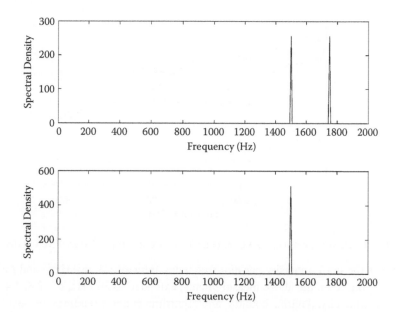

FIGURE 3.2 Cosine signals sampled at f_s = 4000 Hz with M = 512 samples. Upper plot: the signal is the sum of a cosine at 1500 Hz and a cosine at 1750 Hz. Lower plot: the signal is the sum of a cosine at 1500 Hz and a cosine at 2500 Hz.

tribute noise of about 0.1 nV $Hz^{-1/2}$ $\Omega^{-1/2}$. This means that the resistor noise can be comparable to op-amp noise if the input resistors are 100 Ω.

A readily obtainable low-noise microphone, such as the Knowles MR8540, has a self-noise SPL of 30 dB for a 1 kHz bandwidth, or an equivalent input *RMS* acoustic pressure of 6×10^{-4} Pa. Given a sensitivity of -62 dB relative to 1 V/0.1 Pa, its noise output is $(10^{-62/20}/0.1)$ $(6 \times 10^{-4})/(1000^{1/2})$ = 160 nVrms/$Hz^{-1/2}$. Hence microphone self-noise can be expected to be a dominant system noise source.

Background acoustic noise can vary hugely with site, with airports and roadsides being particularly noisy. Acoustic remote-sensing systems generally use very narrow band-pass filters (perhaps 100 Hz wide), so most pure tones, such as from birds, are excluded, and much of the broadband acoustic noise is also greatly reduced. It is important, if the dynamic range of the instrumentation is limited, to band-pass filter at an early stage in the amplifier chain, so as to remove such noise components before they saturate the circuits and cause distortion. Figure 3.3 shows some measured background noise levels.

These and similar measurements by others suggest a simple power-law dependence on frequency of the form

$$N = N_0 \left(\frac{f}{f_0} \right)^{-q},$$

(3.14)

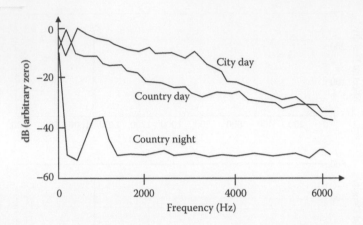

FIGURE 3.3   Power spectra of background acoustic noise at typical sites, given in dB.

where N is the noise intensity per unit frequency interval (W m^{-2} Hz^{-1}) and f is the frequency. Based on the above measurements, extended to 20 kHz, $q \sim 2.8$, 1.4, and 0.5 for daytime city, daytime country, and nighttime country readings, respectively.

3.4   REFLECTION AND REFRACTION

When a sound wave meets an interface where the sound speed changes, some energy is reflected and some continues across the interface but with a change in direction. This can be visualized using the Huygens principle, which states that each point on a wavefront acts like a point source of spherical wavelets, and taking the tangential curve to the wavelets after a short time gives the position of the propagated wavefront. Imagine a plane wavefront meeting a horizontal interface between medium 1 and medium 2 at an angle of incidence θ_i as shown in Figure 3.4. From the construction in medium 1, it can be seen that the triangles ABC and CDA are identical and that the angle of incidence is equal to the angle of reflection.

$$\theta_r = \theta_i. \tag{3.15}$$

Also

$$AC = \frac{BC}{\sin \theta_i} = \frac{AE}{\sin \theta_r}$$

or

$$\frac{c_2}{\sin \theta_r} = \frac{c_1}{\sin \theta_i}, \tag{3.16}$$

which is Snell's law.

Generally, for sound traveling through the air, there is no distinct interface but rather a continuous change in sound speed due to a temperature gradient or wind

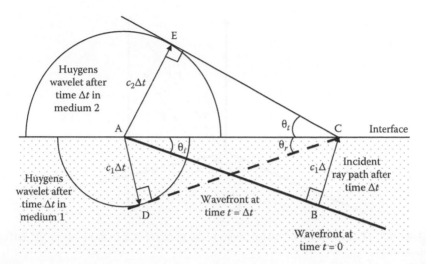

FIGURE 3.4   A wavefront AB incident at an angle θ_i at time $t = 0$ and meeting an interface between medium 1 and medium 2 at point A. After a time Δt the ray from point B meets the interface at C and the Huygens wavelet for the backward, reflected, wave has reached point D. The line CD defines the reflected wavefront. The Huygens wavelet in medium 2 is shown traveling at speed $c2 > c1$, and the transmitted, or refracted, ray reaches point E in time Δt. The line CE defines the refracted wavefront.

shear. In the case where the atmosphere is horizontally uniform and the vertical sound speed gradient dc/dz is constant,

$$\frac{c_0}{\sin \theta_0} = \frac{c}{\sin \theta} = c\sqrt{1 + \cot^2 \theta} = \left(c_0 + \frac{dc}{dz} z \right)\sqrt{1 + \left(\frac{dz}{dx} \right)^2}$$

or, upon integrating,

$$(x - x_0)^2 + (z - z_0)^2 = r^2, \qquad x_0 = \frac{c_0}{(dc / dz)\tan \theta_0},$$

$$z_0 = -\frac{c_0}{dc / dz}, \qquad r = \frac{c_0}{(dc / dz)\sin \theta_0}. \qquad (3.17)$$

The sound propagation path is therefore along a circular arc of radius r and center (x_0, z_0). However, the curvature is usually very small. For example, if $c_0 = 340$ m s^{-1} and $\theta_0 = \pi/10$, the radius of curvature for an adiabatic lapse rate is 67000 km. So in most situations involving acoustic remote-sensing, refraction can be ignored.

The fraction of incident energy reflected from the atmosphere is extremely small (see later) but for most other surfaces and for the frequency ranges typically used for acoustic remote sensing, virtually all sound is reflected. This is an important consideration for siting of acoustic remote-sensing instruments, since even reflections from very distant solid objects can masquerade as genuine atmospheric reflections (known as "clutter" or "fixed echoes").

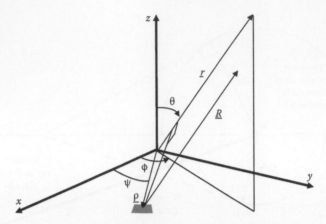

FIGURE 3.5 The geometry of contributions to the pressure at $\underline{R}$ from points on the surface of an antenna.

In the case of acoustic travel-time tomography where the propagation path is at a few meters above the ground, ground reflections can be a major consideration. In this case, the reflection from the ground can combine out of phase with the direct line-of-sight signal, causing a much reduced signal amplitude. For this reason, as discussed further later, continuous encoded-signal systems may experience difficulties and short pulses are generally used.

3.5   DIFFRACTION

SODARs and RASS use antennas, which make the source and the receiver extend over a larger area. The acoustic pressure at some point $\underline{R}$ is the sum of all the pressure contributions from small areas $\rho\,d\psi\,d\rho$ on the antenna surface, as shown in Figure 3.5. The pressure contribution at $\underline{R}$ from an element at position $\underline{\rho}$ will be proportional to the element's area, giving

$$dp = \frac{A(\rho)}{R}e^{j(\omega t - kR)}(\rho\,d\psi)\,d\rho$$

allowing for spherical spreading, the phase at R compared with the phase at r, and an amplitude A varying with position on the antenna.

Also, $R = r - \rho$ so for distances $R \gg \rho$,

$$R = \sqrt{\underline{R}\cdot\underline{R}} = \sqrt{r^2 - 2\underline{r}\cdot\underline{\rho} + \rho^2} \approx r - \rho\sin\theta\cos(\psi + \phi)$$

and, if the antenna gain is uniform across the antenna,

$$p = \frac{2\pi A e^{j(\omega t - kr)}}{r}\int_0^a\left[\frac{1}{\pi}\int_0^\pi e^{jk\rho\sin\theta\cos(\psi+\phi)}\,d(\psi+\phi)\right]\rho\,d\rho,$$

where a is the antenna radius. The integral in the square brackets is the Bessel function $J_0(k\rho \sin \theta)$ and

$$\int_0^x J_0(x)x\,dx = xJ_1(x),$$

so

$$p = \frac{A\,e^{j(\omega t - kr)}}{r}(\pi a^2)\left[\frac{2J_1(ka\sin\theta)}{ka\sin\theta}\right]. \tag{3.18}$$

The oscillatory nature of the last term in square brackets is known as a *diffraction pattern*. It arises because the antenna is not producing a plane wave, but has finite width. This pattern is shown in Figure 3.6. Bands of energy occur at periodic values of θ, which are known as *side lobes*. Depending on the ratio of radius a to wavelength λ, these side lobes can send acoustic power out at low angles and cause reception of echoes from buildings or other structures nearby. It can be seen that the first zero crossing is at $ka \sin \theta = 3.83$, so, for example, if a dish of radius 1 m is used at a wavelength of 0.1 m, then the first zero occurs at $\theta = \sin^{-1}(3.83/62.83) = 3.5°$ and the resulting beam is 7° in width.

Similar oscillating diffraction patterns occur whenever sound impinges on an edge.

3.6   DOPPLER SHIFT

Doppler shift is a change in the frequency of a signal caused by a moving source or target. Imagine a target (a patch of turbulence, for example) moving in the direction of propagation at a speed u and the speed of sound is c, as in Figure 3.7.

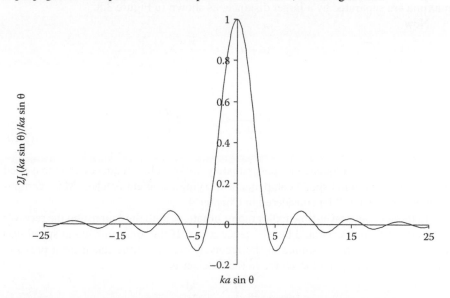

FIGURE 3.6   The diffraction pattern from a circular aperture of uniform gain.

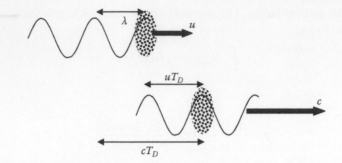

FIGURE 3.7 A turbulent patch moving with speed u in the direction of sound propagation. The lower plot shows the distance moved by the patch in time T_D, and the distance moved by the acoustic pressure wave in the same time.

At time $t = 0$, an acoustic pressure maximum is at the target, and the next pressure maximum is a distance λ away. If this next pressure maximum reaches the target at $t = T_D$, the target has moved a distance uT_D and the pressure maximum has moved a distance $cT_D = \lambda + uT_D$. So the period between two maxima at the target is $T_D = \lambda/(c-u)$. The frequency of the sound at the target is therefore

$$f_D = \frac{1}{T_D} = \frac{c-u}{\lambda} = \frac{c}{\lambda}\left(1-\frac{u}{c}\right) = f\left(1-\frac{u}{c}\right). \tag{3.19}$$

The *Doppler frequency* f_D is less than the transmitted frequency, as sensed by the target.

If the sound is reflected by the target back toward the source, successive pressure maxima are separated by a larger distance, as shown in Figure 3.8.

Now

$$\lambda_D = (c+u)T_D = \frac{c+u}{c-u}\lambda,$$

so

$$f_D = f\frac{c-u}{c+u} \approx f\left(1-2\frac{u}{c}\right). \tag{3.20}$$

The change in frequency is approximately $2(u/c)f$. This frequency change is used to determine the wind speed components carrying turbulent patches. More complicated geometries will be considered in Chapter 4.

In the acoustic travel-time tomography situation, both the source and the receiver are stationary, and separated by a distance $x = X$. If the air is moving at speed $u(x)$ along the line from the source to the receiver, then the time taken for a pressure maximum to move from the source to the receiver is

$$t_{\text{downwind}} = \int_0^X \frac{dx}{c(x) + u(x)} \tag{3.21}$$

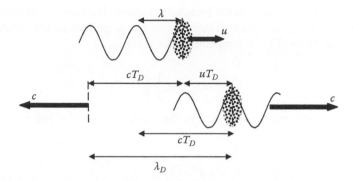

FIGURE 3.8 Reflection of sound from a target moving in the direction of sound propagation. The dashed lines show positions of reflected pressure maxima at a time T_D after the first pressure maximum reaches the target patch.

and in the opposite direction

$$t_{\text{upwind}} = \int_0^X \frac{dx}{c(x) - u(x)}, \tag{3.22}$$

where both wind speed and sound speed can, in general, vary along the path. These times are identical for successive pressure maxima so *there is no Doppler shift*. However, the downwind and upwind travel times can distinguish temperature variations (changes in c) from wind speed variations (changes in u) since $u \ll c$ and

$$t_{\text{upwind}} - t_{\text{downwind}} = \int_0^X \frac{dx}{c(1 - u/c)} - \int_0^X \frac{dx}{c(1 + u/c)} \approx \int_0^X \left(1 + \frac{u}{c}\right) \frac{dx}{c} -$$

$$\int_0^X \left(1 - \frac{u}{c}\right) \frac{dx}{c} \approx 2 \int_0^X \frac{u \, dx}{c^2}, \qquad t_{\text{upwind}} + t_{\text{downwind}} \approx 2 \int_0^X \frac{dx}{c}. \tag{3.23}$$

3.7 SCATTERING

Scattering of sound by turbulence has been very thoroughly investigated theoretically (Tatarskii, 1961; Ostashev, 1997). Here we give a more intuitive description, together with some new results relating to SODARs.

3.7.1 Scattering from Turbulence

Scattering occurs when an object with a sound speed different from air causes rays from the wavefront to deviate into many directions. In the case of scattering from turbulent temperature fluctuations, there are many randomly placed and randomly sized scatterers, each having a density very slightly different from the average air density. Scattering can also be caused by the random motion of the turbulent patches since this too causes a change in the local sound speed. The strength of such scattering (how much energy is deflected) depends on the magnitude of the variations c' in sound speed. From (3.2) for temperature fluctuations T'

$$\frac{c'}{T'} \approx \frac{dc}{dT} = \frac{1}{2}\frac{c}{T}. \tag{3.24}$$

Generally sound speed variations are expressed as refractive index fluctuations of magnitude

$$n' = \frac{c'}{c}. \tag{3.25}$$

For a fluctuation $\underline{V}'$ in the vector wind, the sound speed fluctuation depends on the direction of $\underline{V}'$ compared to the direction of propagation $\hat{k}$ of the sound ($\hat{k}$ is a unit vector). The combination of temperature and velocity fluctuations gives refractive index fluctuations

$$n' = \frac{V' \cdot \hat{k}}{c} + \frac{T'}{2T}. \tag{3.26}$$

The following chiefly relates to SODARs since they obtain a signal through reflections from turbulence. The SODAR beam and pulse duration τ define a volume that contains refractive index fluctuations n' continuously varying in strength and spatial scale. Scattering from this volume is a three-dimensional problem, but the general ideas can be more easily understood by considering propagation and scattering of sound in just the vertical, z, direction. Consider two layers spaced by l and having refractive index fluctuations $n'(z)$ and $n'(z+l)$ as shown in Figure 3.9.

The amplitude from a single fluctuation is proportional to n'. The sound intensity I is proportional to the *square* of the sum of all the individual scattered amplitudes and contains terms like

$$I \propto \overline{n'(z)n'(z+l)}e^{-jk2l}. \tag{3.27}$$

The bar over the refractive index product means that the fluctuations have been averaged over time, and the exponential term accounts for the path difference of $2l$ for backscatter from the patches at z and at $z+l$ (i.e., this is a phase term). The wavelength of the transmitted sound is λ and the corresponding wavenumber is $k = 2\pi/\lambda$. Integrations must be performed over the z range $\pm c\tau/2$ of the SODAR scattering volume, and over the separations l, thus

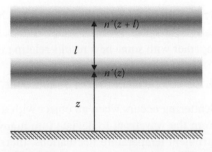

FIGURE 3.9 The geometry of two scattering layers.

$$I \propto \int_l \left[\int_z \overline{n'(z)n'(z+l)}\, dz \right] e^{-jk2l}\, dl. \tag{3.28}$$

The term in square brackets is called the spatial *autocorrelation* function of fluctuations n'. This decreases with increasing separation l between turbulent layers as they become increasing uncorrelated and $n'(z)$ and $n'(z+l)$ are less likely to increase or decrease together. The autocorrelation of refractive index fluctuations therefore contains information about *spatial scales* of turbulence. This information could also be expressed in terms of spatial frequencies by taking the *Fourier transform* of this autocorrelation function. The Wiener–Khinchine theorem shows that the Fourier transform of the autocorrelation function is the *power spectrum*, Φ_n, of n', or

$$\Phi_n(\kappa) = \int_l \left[\int_z \overline{n'(z)n'(z+l)}\, dz \right] e^{-j\kappa l}\, dl. \tag{3.29}$$

Using the inverse Fourier transform, Eq. (3.28) can be written in the form

$$I \propto \int_l \left[\int_\kappa \Phi_n(\kappa)e^{j\kappa l}\, d\kappa \right] e^{-jk2l}\, dl, \tag{3.30}$$

where κ is a spatial wavenumber, $\kappa = 2\pi/d$, for refractive index fluctuations of size d as in (2.13) of Chapter 2. This can be rearranged

$$I \propto \int_\kappa \Phi_n(\kappa) \left[\int_{-c\tau/2}^{c\tau/2} e^{-j(2k-\kappa)l}\, dl \right] d\kappa. \tag{3.31}$$

The term in square brackets is the Fourier transform of a rectangular function of length $c\tau$

$$\int_{-c\tau/2}^{c\tau/2} e^{-j(2k-\kappa)l}\, dl = \frac{c\tau}{2} \frac{\sin[(2k-\kappa)c\tau/2]}{(2k-\kappa)c\tau/2}, \tag{3.32}$$

which has the shape shown in Figure 3.10.

The zero crossings occur at

$$2k - \kappa = \pm \frac{4\pi}{c\tau}. \tag{3.33}$$

Typical values for a SODAR are $c\tau = 8.5$ m, or $4\pi/c\tau = 0.7$ m^{-1}, and $\lambda = 0.08$ m, or $k = 80$ m^{-1}. So κ is very nearly equal to $2k$, and the term in square brackets in Eq. (3.32) is like a delta-function $\delta(\kappa - 2k \pm 4\pi/c\tau)$, which has the property

$$\int_\kappa \Phi_n(\kappa)\delta\left(\kappa - 2k \pm \frac{4\pi}{c\tau}\right) d\kappa = \Phi_n\left(2k \pm \frac{4\pi}{c\tau}\right). \tag{3.34}$$

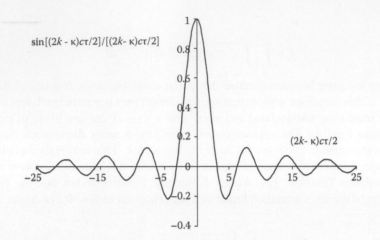

FIGURE 3.10 The Fourier transform of a rectangular function of length $c\tau/2$.

The result is that

$$I \propto \Phi_n\left(2k \pm \frac{4\pi}{c\tau}\right). \tag{3.35}$$

This means that the reflected sound intensity depends on the strength of refractive index fluctuations having spatial wavenumbers of *twice* the wavenumber of the transmitted sound. This can be interpreted simply as follows. The scattered sound is very weak, but scatterings separated vertically by $d = \lambda/2$ will add in phase (there is no π phase change on reflection for soft scattering). This means that $\kappa = 2\pi/d = 2\pi/(\lambda/2) = 2k$ as predicted by (3.35).

The above simplified analysis applies for scattering directly back to a "monostatic" SODAR which has speakers and microphones located at the same place. The situation is more complicated for "bistatic" SODARs for which the scattering angle is not 180°. Figure 3.11 shows scattering through an angle β from two turbulent patches at positions A and B which are in layers separated by l. As in the case of 180° scattering, the incident ray (shown as a dark line) is at an angle $\beta/2$ to the layers. The extra path length for sound scattered from B, compared to that scattered from A, is distance ABC where

$$\text{ABC} = l\frac{1 + \sin(\beta - \pi/2)}{\sin(\beta/2)} = 2l\sin\frac{\beta}{2}. \tag{3.36}$$

When the path difference ABC equals λ, a strong signal results because the scattered waves are then in phase.

The intensity in the general case is therefore proportional to $\Phi_n(2k\sin(\beta/2) \pm 4\pi/c\tau)$ where $c\tau$ is a typical dimension of the scattering volume. For example, if the frequency is 5100 Hz, then for backscatter $2k \sim 2\pi\,60$ m^{-1} and the volume correction term $4\pi/c\tau$ is small providing $c\tau \gg 1/60$ m. The theory predicts very tight dependence on the *Bragg wavenumber* $2k\sin(\beta/2)$. Also, it is often con-

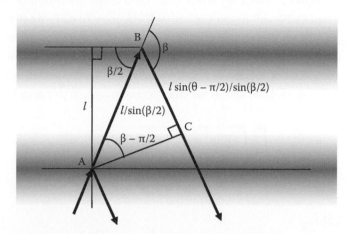

FIGURE 3.11 The geometry of scattering for the general bi-static SODAR case, where scattering is through angle β from layers separated by distance l.

venient to think of the turbulent wavenumber as being a measure of turbulent eddy size, since eddies spaced by κ can be thought of as having dimensions of approximately $2\pi/\kappa$. This means that a SODAR of wavelength λ, set up so that scattering is through an angle β, will record echo information about the intensity of refractive index fluctuations of size

$$d = \frac{\lambda}{2\sin(\beta/2) \pm \lambda/c\tau}. \tag{3.37}$$

As an example, if the wavelength is 0.1 m (typical of a SODAR operating at 3.4 kHz) and the pulse duration is 50 ms, then $c\tau = 10$ m and for monostatic sounding (with $\beta = 180°$), $d = 0.0498$ to 0.0503 m.

It should not be thought from the above that all the turbulent patches of size d somehow "line up" to give a resonant back-scatter in the manner of regular Bragg scattering from a crystal. Rather, the incident wavetrain picks out those spatial Fourier components which give the strongest combined reflection. Physically, we can imagine a spatial arrangement of fluctuations multiplied by a sine wave. Averaging over the record length then gives a measure of the combined strength of reflections. But this process is identical to finding the coefficients in a Fourier series, which can be generalized via Fourier transforms.

3.7.2 INTENSITY IN TERMS OF STRUCTURE FUNCTION PARAMETERS

From the previous section it is clear that the scattered acoustic intensity received by a SODAR is proportional to the power spectrum, $\Phi_n(2k\sin(\beta/2))$, of refractive index fluctuations at spatial wavenumbers close to $\Phi_n = 2k\sin(\beta/2$. What is the connection between Φ_n and the turbulence parameters such as C_V^2 and C_T^2

discussed in Chapter 2? Since Φ_n arises from the Fourier transform of terms like $\overline{n'(z)n'(z+l)}$ in (3.29) and from (3.26) $n' = \underline{V}' \cdot \hat{\underline{k}}/c + T'/2T$, we would expect Φ_n to be the sum of terms containing the power spectra of the autocorrelation functions of $[\underline{V}'(z) \cdot \hat{\underline{k}}][\underline{V}'(z+l) \cdot \hat{\underline{k}}]$ and $\overline{T'(z)T'(z+l)}$. In general, there would also be cross-terms in $V'T'$ but it is usually assumed that the fluctuations in velocity and temperature are uncorrelated and so the cross-terms vanish. Writing, as in (3.29),

$$\Phi_V(\kappa) = \int_l \left[\int_z \overline{[\underline{V}'(z) \cdot \hat{\underline{k}}][\underline{V}'(z+l) \cdot \hat{\underline{k}}]} dz \right] e^{-j\kappa l} \, dl,$$

$$\Phi_T(\kappa) = \int_l \left[\int_z \overline{T'(z)T'(z+l)} dz \right] e^{-j\kappa l} \, dl, \tag{3.38}$$

then from (3.26)

$$\Phi_n\left(2k\sin\frac{\beta}{2}\right) = \cos^2\beta \left[\frac{\Phi_T(2k\sin(\beta/2))}{4T^2} + \frac{1}{4\pi}\cos^2\left(\frac{\beta}{2}\right)\frac{\Phi_V(2k\sin(\beta/2))}{c^2} \right]. \tag{3.39}$$

The dot product in $\underline{V}'(z) \cdot \hat{\underline{k}}$ gives a cosine of the angle between $\underline{V}'(z)$ and $\hat{\underline{k}}$. Averaging over the square of such terms for a total scattering angle of β gives the $(1/4\pi)\cos^2(\beta/2)$ term. The $\cos^2\beta$ term allows for the fact that the "reflections" from the two turbulent patches at A and B in Figure 3.11 are not diffuse.

From Chapter 2, $\Phi_V(\kappa) = 0.76C_V^2\kappa^{-5/3}$ and $\Phi_T = 0.033C_T^2\kappa^{-5/3}$, so the intensity from combined temperature and velocity fluctuations is

$$I \propto \cos^2\beta \left[0.033\frac{C_T^2}{T^2} + 0.76\cos^2\left(\frac{\beta}{2}\right)\frac{C_V^2}{\pi c^2} \right]. \tag{3.40}$$

This has the very interesting property that acoustic backscatter (with $\beta = 180°$) from turbulence depends *only* on the temperature fluctuations. Monostatic SODARs are therefore insensitive to mechanical turbulence and give very low signals in near-neutral conditions when there is little temperature contrast.

3.7.3 SCATTERING FROM RAIN

Rain can be a significant source of acoustic echoes for SODARs. Each drop acts as an individual scatterer and, since a drop diameter D is small compared with the wavelength λ, Rayleigh acoustic scattering is a good approximation. In this approximation, the entire drop volume experiences the same acoustic pressure at any one time, and all parts of the drop radiate acoustic energy in phase, so the scattered amplitude is proportional to the drop volume, or to diameter D cubed. The scattered intensity I_s is proportional to the incident intensity I_i, to the square of the scattered amplitude, and also decreases with distance r squared (i.e., spherical spreading). Assume that there is also a dependence on wavelength λ to the power q, so

$$\frac{I_s}{I_i} = A\frac{D^6}{r^2}\lambda^q,$$

where A is a dimensionless constant. Since the left side is dimensionless, $q = -4$. When a full solution of the acoustic wave equation is done, each drop acts as if there is an equivalent plane area scattering all incident energy. This equivalent cross-section area, per unit solid angle into which the sound is scattered, is called the *differential scattering cross-section* σ_R, and is given by

$$\sigma_R = \frac{\pi^5}{36}(2+3\cos\beta)^2\frac{D^6}{\lambda^4}. \tag{3.41}$$

Figure 3.12 shows the angular dependence of this rain scattering, together with the angular dependence of scattering by turbulent temperature and velocity fluctuations. The relative acoustic intensities from the three scattering mechanisms at any angle depend on the magnitudes of rainfall intensity R, C_T^2, and C_V^2, and since generally $C_V^2 \gg C_T^2$ the scattered intensity from velocity fluctuations can exceed those from temperature fluctuations within a few degrees of the back-scatter direction.

It is worth noting that the peak scattered intensity from turbulence for bi-static SODARs occurs at an angle β given by

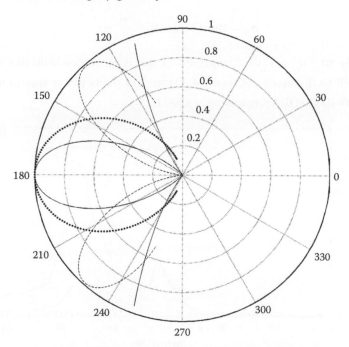

FIGURE 3.12 The angular dependence of scattering in the backward direction: from rain (solid line), from temperature fluctuations (dotted line), and from velocity fluctuations (dashed line). Arbitrary scaling has been used.

$$\frac{d}{d\beta}\left(\cos^2\beta\cos^2\frac{\beta}{2}\right)=0$$

or $\cos\beta = -2/3$, which is *exactly* where the scattering from rain has a minimum. Bistatic SODARs are therefore less susceptible to rain clutter.

The total differential acoustic scattering cross-section for rain includes scattering from drops of all diameters, weighted according to the numbers $n_D(D)\,dD$ of drops per unit volume having diameters between D and dD. A commonly assumed probability distribution for raindrop diameters is the Marshall–Palmer distribution (Marshall and Palmer, 1948)

$$n_D(D) = n_0\,e^{-\Lambda D}, \tag{3.42}$$

where $n_0 = 8 \times 10^6$ m^{-4} for drop diameters D in m and $\Lambda = 4100 R^{-0.21}$ m^{-1} for rainfall intensity R in mm h^{-1}. Integrated over all drop sizes (in practice limited to about 6 mm maximum diameter), there are about 2000 drops per m^3 for $R = 1$ mm h^{-1}. The integrated differential scattering cross-section is

$$\sigma_R = 20\pi^5(2+3\cos\beta)^2\frac{n_0}{\lambda^4\Lambda^7} \tag{3.43}$$

or typically 10^{-11} m^2 per m^2. Comparison of this cross-section with that from turbulence will be discussed in Chapter 4. Figure 3.13 shows power spectra measured using an AV4000 SODAR operating at 4500 Hz.

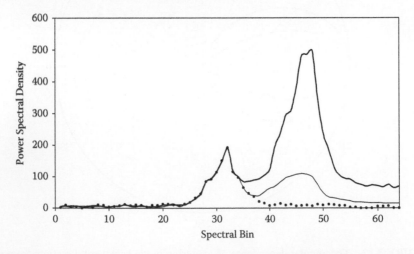

FIGURE 3.13   Power spectra recorded from a vertically pointing SODAR with: no rain (dotted curve), 5 to 10 mm/h (thin line), and greater than 10 mm/h rain (heavy line).

3.8   ATTENUATION

3.8.1   Losses Due to Spherical Spreading

Equation (3.4) describes a *plane wave* in which there is no variation in intensity in the x and y directions. In practice, the sound spreads out from a localized source such as a SODAR. If the transmitted power, P_T, is spread out evenly into a sphere of radius r, then the intensity at distance r from the source would be

$$I = \frac{\text{power}}{\text{area}} = \frac{P_T}{4\pi r^2}. \tag{3.44}$$

The intensity clearly decreases with range squared. From (3.9), this is equivalent to a loss of $20 \log_{10} 2 = 6$ dB for every doubling of range. This is one reason for the limited range of SODAR and RASS instruments.

3.8.2   Losses Due to Absorption

When sound travels a small distance through air, the intensity I decreases a small amount ΔI due to absorption losses. The amount of intensity decrease depends on the distance Δx traveled and also depends on the initial energy, so $\Delta I \propto -I\Delta x$, or

$$\frac{dI}{I} = -\alpha \, dx, \tag{3.45}$$

where α is the *absorption coefficient*. If α does not vary along the sound path, then integrating gives

$$I = I_0 \, e^{-\alpha x}. \tag{3.46}$$

The absorption coefficient α is the sum of classical absorption, α_c, and molecular absorption, α_m. Classical absorption is due to each small volume of air being compressed and stretched by the sound pressure along the direction of propagation, and so causing a shape change or shear, which is resisted by viscous forces. The energy loss per cycle is proportional to the shear, which is proportional to the size of the volume affected or to the wavelength λ. The energy loss per second is therefore proportional to the square of frequency, f^2. This means that energy loss per unit length is also proportional to f^2.

Molecular absorption is due to transfer of a molecule's energy out of the translational motion and into vibration or rotation of the molecule. For dry air consisting of N_2 and O_2 molecules, any extra energy transferred to rotation during a sound pressure impulse is transferred back into translational energy in a very short time compared to the sound wave period. For these molecules, the "relaxation time" for rotational modes is very short. On the other hand, excess energy is not transferred efficiently to vibrational modes because their relaxation time is very long. So dry air does not have much molecular absorption. However, when water vapor molecules are present, the transfer of energy to vibrational modes in N_2 and O_2 occurs in a very much shorter time through collisions with H_2O. But at high humidities, O_2 and N_2 molecules are fully excited in their vibration mode without acoustically enhanced collisions, and there is again little extra energy taken out of the pressure wave. Absorption also depends on temperature and pressure since these affect collisions. Absorption in dB m^{-1} is a very complicated formula (Salomons, 2001)

$$\alpha = \frac{8.686 f^2}{\sqrt{\mu}} \left\{ 1.84 \times 10^{-11} \frac{p_0}{p} + \left[\frac{0.1068 f_{N_2} \exp(-3352/T)}{f_{N_2}^2 + f^2} + \frac{0.01275 f_{O_2} \exp(-2239.1/T)}{f_{N_2}^2 + f^2} \right] \mu^3 \right\},$$

$$f_{N_2} = \frac{p}{p_0} [9 + 280 q e^{-4.17(\mu^3 - 1)}] \sqrt{\mu},$$

$$f_{O_2} = \frac{p}{p_0} \left(24 + 40400 q \frac{0.02 + q}{0.391 + q} \right),$$

$$\ln\left(\frac{p}{p_0} \frac{q}{r} \right) = 10.6267 - 15.7372 \left(\frac{T_0}{T} \right)^{1.261},$$

$$\mu = \frac{T_1}{T},$$

(3.47)

where f is the frequency in Hz, T the temperature in K, p the pressure in kPa, r the relative humidity in %, and the constants are $p_0 = 101.325$ kPa, $T_0 = 273.16$ K, and $T_1 = 293.15$ K. The result for several values of r is shown in Figure 3.14 at $T = 10°C$ (upper plot) and $T = 20°C$ (lower plot) and with $p = 101.3$ kPa. The slope is close to that of f^2 over the usual range of SODAR frequencies from 1.5 to 6 kHz and humidity ranging from 20% to 100%. A simple approximation for this range, with r in %, f in kHz, and air temperature T_C in the range 10 to 20°C is

$$\alpha = 0.0018 f^2 \frac{p_0}{p} [1 + 10 e^{-r/(22 - 0.6 T_C)}] \text{ dB m}^{-1}. \qquad (3.48)$$

Absorption coefficients based on this approximation are also shown in Figure 3.14 for $f = 3$ kHz. Note that absorption in dB m^{-1} is equal to $10 \log_{10} e$ times the absorption in m^{-1}. A 2 kHz SODAR would lose an extra 40 dB due to absorption when the humidity is 50% and temperature 10°C, and 20 dB due to beam spreading between 100 m and 1 km range. Higher frequency SODARs will have more limited range due to absorption.

3.8.3 LOSSES DUE TO SCATTERING OUT OF THE BEAM

Similar to rain, turbulence acts as if there is an extremely small equivalent plane area scattering all incident energy within a beam. This equivalent cross-section area, per unit volume and per unit solid angle into which the sound is scattered, is called the differential scattering cross-section σ_s. In Chapter 4, we will find that

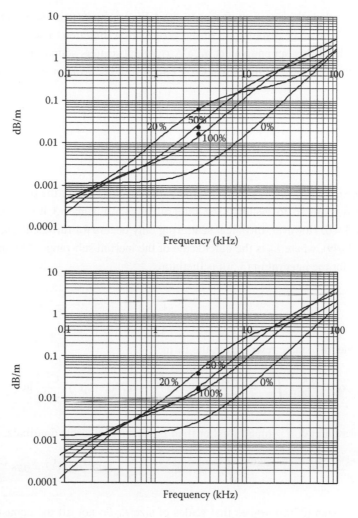

FIGURE 3.14 Atmospheric absorption of sound in dB m^{-1} as a function of sound frequency. Temperature $T = 10°C$ (upper plot) and $T = 20°C$ (lower plot) and with $p = 101.3$ kPa. The individual points plotted are from Eq. (3.48).

$$\sigma_s = \frac{1}{8} \frac{k^4 \cos^2\beta}{\kappa^{11/3}} \left[0.033\frac{C_T^2}{T^2} + 0.76\cos^2\left(\frac{\beta}{2}\right)\frac{C_V^2}{\pi c^2} \right] =$$

$$\frac{k^{1/3}}{2^{20/3}} \frac{\cos^2\beta}{\left(\sin\frac{\beta}{2}\right)^{11/3}} \left[0.033\frac{C_T^2}{T^2} + 0.76\cos^2\left(\frac{\beta}{2}\right)\frac{C_V^2}{\pi c^2} \right].$$

$$(3.49)$$

Assume that the incident intensity is I_i W m^{-2} in a volume of cross-section area A and small distance dz in the propagation direction. The total incident power is I_iA. Of this, an equivalent area A σ_s dz $d\Omega$ scatters sound into solid angle $d\Omega = 2\pi \sin \beta \, d\beta$. Integrating over all solid angles, the equivalent scattering area is

$2\pi \, dz A \int \sigma_s(\beta) \sin\beta \, d\beta$, giving a loss in ongoing intensity of $dI_i = -\alpha_e I_i \, dz$ where

$$\alpha_e = 2\pi \int \sigma_s(\beta) \sin\beta \, d\beta \qquad (3.50)$$

is the *excess attenuation* due to sound being scattered out of the beam.

Care is required to evaluate the integral in (3.50), since (3.49) suggests that $\sigma_s = \infty$ at $\beta = 0$. However, turbulent scattering only occurs for turbulent scales up to $d = L_0$ in (3.49), where L_0 is the outer scale of the inertial sub-range. The minimum value of β for the integral in (3.50) is therefore

$$\beta_{\min} = 2\sin^{-1}\left(\frac{\lambda}{2L_0}\right). \qquad (3.51)$$

The integral can be rewritten in the form

$$\alpha_e = \frac{\pi k^{1/3}}{2^{11/3}} \int_{\lambda/2L_0}^{1} \frac{(1-2\mu^2)^2}{\mu^{8/3}} \left[0.033 \frac{C_T^2}{T^2} + 0.76(1-2\mu^2)^2 \frac{C_V^2}{\pi c^2} \right] d\mu.$$

The wavelength λ is much smaller than L_0, so small μ terms dominate, and

$$\alpha_e \approx \frac{\pi k^{1/3}}{2^{11/3}} \left[0.033 \frac{C_T^2}{T^2} + 0.76 \frac{C_V^2}{\pi c^2} \right] \int_{\lambda/2L_0}^{1} \mu^{-8/3} \, d\mu \approx \frac{3}{5\pi^{2/3} 2^{11/3}} k^2 L_0^{5/3} \left(0.033 \frac{C_T^2}{T^2} + 0.76 \frac{C_V^2}{\pi c^2} \right).$$

Pan-Naixian (2003) argues that scales of size L_0 do not fill the acoustic beam for conventional SODARs, and so the appropriate limiting turbulent element size is $2z\Delta\theta$, where $\Delta\theta$ is the beam angular half-width (typically 5°). This half-width is inversely proportional to transmitting frequency for an antenna of fixed radius (as will be seen in Chapter 4), having the form $\Delta\theta = 1.62/ka$. Also, the term in C_V^2 dominates over the term in C_T^2, so the result is

$$\alpha_e \approx 0.04 k^{1/3} \frac{C_V^2}{c^2} \left(\frac{z}{a}\right)^{5/3}. \qquad (3.52)$$

Pan-Naixian produces evidence for this dependence on frequency. For a transmit frequency $f_T = 3400$ Hz, $c = 340$ m s^{-1}, $\lambda = 0.1$ m, $k = 63$ m^{-1}, $a = 0.6$ m, $z = 200$ m, and $C_V^2 = 0.01$ m$^{4/3}$ s^{-2}, we find $\alpha_e = 2 \times 10^{-4}$ m^{-1}. Further corrections for non-uniform intensity across the SODAR beam do not change the conclusion that $\alpha_e \ll \alpha$.

3.9 SOUND PROPAGATION HORIZONTALLY

Acoustic travel-time tomography also relies on sound propagating through the atmosphere, although in this case propagation is horizontal and close to the surface (Raabe et al., 2001; Ziemann et al., 2001). Measurements are conducted using multiple paths, each of which is between an acoustic source and an acoustic receiver. The *only* measurement made is the time taken for the sound to travel a path.

There is no Doppler shift since the travel time between the source and the receiver is the same for each acoustic wavefront: if successive wavefronts are period T apart initially, they also arrive period T apart.

The time taken to travel a fixed distance r is

$$t(r) = \int_0^r \frac{dr}{c(r)}, \tag{3.53}$$

where $1/c$ is often called the *slowness*. The temperature and wind effects on c are generally small, so (3.26) can be used and

$$\frac{1}{c(\underline{r})} \approx \frac{1}{c(0)} \left[1 + \frac{V \cdot \hat{\underline{r}}}{c(0)} + \frac{T(\underline{r}) - T(0)}{2T(0)} \right]^{-1} \approx \frac{1}{c(0)} \left[1 - \frac{V \cdot \hat{\underline{r}}}{c(0)} - \frac{T(\underline{r}) - T(0)}{2T(0)} \right] \tag{3.54}$$

giving the travel times in both directions along the path as

$$t(\underline{r}) = \frac{1}{c(0)} \left[r - \frac{1}{c(0)} \int_0^r (\underline{V} \cdot \hat{r}) \, dr - \frac{1}{2T(0)} \int_0^r \{ T(\underline{r}) - T(0) \} \, dr \right],$$

$$t(-\underline{r}) = \frac{1}{c(0)} \left[r + \frac{1}{c(0)} \int_0^r (\underline{V} \cdot \hat{r}) \, dr - \frac{1}{2T(0)} \int_0^r \{ T(\underline{r}) - T(0) \} \, dr \right]. \tag{3.55}$$

Consequently

$$\frac{1}{2} [t(\underline{r}) + t(-\underline{r})] = \frac{r}{c(0)} - \frac{1}{c(0)T(0)} \int_0^r [T(\underline{r}) - T(0)] \, dr,$$

$$\frac{1}{2} [t(\underline{r}) + t(-\underline{r})] = -\frac{2}{c^2(0)} \int_0^r (\underline{V} \cdot \hat{r}) \, dr \tag{3.56}$$

gives separate equations for wind speed and temperature along the path. More detail is obtained by having multiple paths and, generally, dividing the 2D sampled area into discrete grids with each grid cell having a constant temperature and wind speed. Each of the M paths is represented by an integral pair as in (3.56) and the measured sums and differences of travel times can be written in the form

$$s_m = \sum_{n=1}^{N} \gamma_{mn} \Delta T_n, \qquad t_m = \sum_{n=1}^{N} (\alpha_{mn} u_n + \beta_{mn} v_n), \tag{3.57}$$

where the weights α_{mn}, β_{mn}, and γ_{mn} are simply the known distances within each of the N cells traversed by this sound path, combined with the other known factors in (3.56). The travel-time pair for each bidirectional path contains information about $3N$ unknowns. If there are N^2 cells, there are $3N^2$ unknowns u_n, v_n, and T_n. This means that at least $3N^2$ paths are required. If there are M sensor sites and each site receives sound from every other site, then there will be a maximum of $M(M-1)$ paths (not all these paths may be useful if the sites are in line at any point). Consequently

$$\text{number of sensor sites} \approx \sqrt{3 \times \text{number of cells}}. \tag{3.58}$$

The result is a set of simultaneous equations, which can be solved for the unknown temperature and velocity field.

There are three main difficulties with the method. The first is to conduct almost simultaneous multiple path transmissions using sound without cross-interference between sensors. Secondly, the accuracy of the results depends critically on being able to precisely time the travel time of the signal. This means using some type of pulse coding. Thirdly, the layout of the various paths has to be designed so that the matrix inversion to find the velocity and temperature field is robust. This last requirement also relates to how the inversion is performed: generally some type of constrained linear inverse method is used (Vecherin et al., 2006; Rogers, 2000).

3.10   SUMMARY

The main features of atmospheric sound propagation relevant to atmospheric acoustic remote sensing are:

1. The sound speed increases by 0.17% per degree increase in air temperature.
2. Sound intensity is generally measured on a logarithmic scale, in dB.
3. Recorded signals are digitized at a rate greater than twice the highest frequency present in the signal. Fourier transforms (specifically the FFT) are used to analyze a signal so that a frequency spectrum is obtained. The spectrum extends to the Nyquist frequency, which is half the sampling frequency.
4. Background acoustic noise decreases with increasing frequency.
5. Refraction, due to a changing propagation speed, acts to change the direction of an acoustic beam toward the region of slower speed.
6. The Doppler frequency of sound reflected from an object which is moving directly away from the source at speed u is a fraction $(1-2u/c)$ of the transmitted frequency. When both the source and the detector are stationary, as in acoustic travel-time tomography, there is no Doppler shift.
7. The main losses of signal strength are due to spherical spreading (power decreases inversely with range) and absorption (power decreases between 0.01 and 0.1 dB m^{-1}).
8. Turbulent scales of $\lambda/[2 \sin(\beta/2)]$ scatter sound of wavelength λ through angle β. The scattered intensity depends on the strength of temperature fluctuations and on the strength of velocity fluctuations, but the angular dependence is different for the two types of turbulent fluctuations. In

particular, backscattered sound ($\beta = 180°$) only derives from temperature fluctuations. The scattered intensity can be expressed in terms of structure function parameters C_V^2 and C_T^2.

9. Scattering of sound by rain can be dominant at higher rainfall intensities. Its angular distribution has a minimum where scattering by velocity fluctuations has a maximum.

10. Travel times of sound along a number of horizontal paths can be used to find the velocity and temperature field within an enclosed 2D space. This method does not involve Doppler shift, but does depend on many sensors and many paths.

REFERENCES

Marshall JS, Palmer WM (1948) The distribution of raindrops with size. J Met 5: 165–166.

Ostashev VE (1997) Acoustics in moving inhomogeneous media. E & FN Spon, London.

Pan-Naixian (2003) Excess attenuation of an acoustic beam by turbulence. J Ac Soc Am 114(6): 3102–3111.

Raabe A, Arnold K et al. (2001) Near surface spatially averaged air temperature and wind speed determined by acoustic travel time tomography. Meteorol Z 10(1): 61–70.

Rogers CD (2000) Inverse methods for atmospheric sounding. World Scientific, London.

Salomons EM (2001) Computational atmospheric acoustics. Kluwer Academic Publisher, Dordrecht.

Tatarskii VI (1961) Wave propagation in a turbulent medium. New York, McGraw Hill, 285 pp.

Vecherin SN, Ostashev VE et al. (2006) Time-dependent stochastic inversion in acoustic travel-time tomography of the atmosphere. J Ac Soc Am 119: 2579–2588.

Ziemann A, Arnold A et al. (2001) Acoustic tomography as a method to identify small-scale land surface characteristics. Acta Acoustica 87: 731–737.

4 Sound Transmission and Reception

The essence of an acoustic remote-sensing system is in generating sound into a well-formed beam which interacts with the atmosphere in a known manner and then detecting that interaction. In Chapter 2 we learned about the nature of the atmosphere into which the sound is projected, and in Chapter 3 the way in which sound travels. In this chapter we describe how to form a beam of sound, how scattered sound is detected, and how systems are designed to optimize retrieval of various atmospheric parameters. The main emphasis of Chapter 4 is on geometry and timing, but details on some of these aspects are left to Chapter 5.

4.1 GEOMETRIC OBJECTIVE OF SODAR DESIGN

The boundary layer atmosphere is often strongly varying in the vertical, but horizontally much more homogeneous. The geometric design objective for vertically profiling instruments is therefore to localize the acoustic power sufficiently in space so that atmospheric properties are obtained from well-defined *height* intervals at a particular time. This means that the vertical resolution has to be defined, typically by using a pulsed transmission. But since sound will spread spherically from the source, height resolution also depends on angular width of the beam transmitted. Here we concentrate on SODAR (SOund Detection And Ranging) systems, for which the acoustic beams are often non-vertical, as shown in Figure 4.1.

Here the pulse duration is τ and the angular width of the acoustic beam is $\pm\Delta\phi$ in azimuth angle and $\pm\Delta\theta$ in zenith angle. From Figure 4.1, the vertical extent of the pulse volume is $\approx c\tau \cos\theta + 2z\Delta\sin\theta\Delta\theta$, which has a term increasing with height z. Taking $c\tau = 20$ m, and $\theta = 20°$, the vertical extent of the pulse volume near the ground is $c\tau \cos\theta = 18.8$ m but, for a beam half-width of $\Delta\theta = 5°$, this increases to 50 m at $z = 500$ m. This emphasizes the need to keep the product $\sin\theta\Delta\theta$ small. Also, if $\Delta\theta$ is too large then the pulse volume will include a wide range of radial velocity, the Doppler spectrum will be wider, and the ability to detect the peak position of the Doppler spectrum, in the presence of noise, will be compromised. But we will see later that the wind velocity component estimates of u and v have errors which depend on $1/\sin\theta$, so it is important that θ not be too small. On the other hand, θ must also not be so large that the volumes sampled by the various SODAR beams which point in different directions, are so spatially separated that their wind components become uncorrelated. The resulting design must therefore be a delicate balance between modest θ values and a narrow beam width $\Delta\theta$. Typical designs have $15° < \theta < 25°$ and $4° < \Delta\theta < 8°$. Obtaining such a small beam width $\Delta\theta$ requires an antenna, since the beam widths of individual speakers are typically much greater. Use of an antenna has the added advantage of increasing the collecting area for echo power.

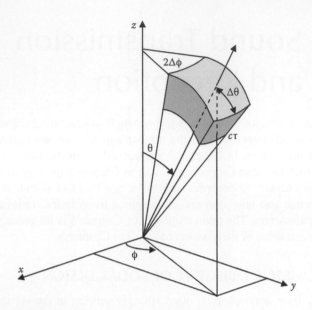

FIGURE 4.1   The pulse volume for a tilted acoustic beam.

4.2   SPEAKERS, HORNS, AND ANTENNAS

4.2.1   SPEAKER POLAR RESPONSE

Figure 4.2 shows several typical speakers. The TOA SC630 is a double re-entrant horn 30-W speaker producing a sound pressure level (SPL) of 113 dB at 1 m and at 1.5 kHz. The FourJay 440-8 "Thundering Mini" is a compact 40-W re-entrant horn speaker with an SPL = 110 dB peaking at 2 kHz. The Motorola KSN1005A is a small piezo-electric horn speaker producing an SPL of 94 dB at 5 kHz. Horn speakers consist of a driver, which includes a diaphragm, and a horn-shaped cone of plastic or metal to efficiently couple energy from the small driver into the atmosphere. Re-entrant horn speakers have the cone split into a backward-facing part connected to the driver and a forward-facing part exiting into the atmosphere; they have the advantage of being more weatherproof and can in many cases be mounted facing upward.

Figure 4.3 shows polar plots of the sound intensity produced by these speakers at selected frequencies. It is clear from these polar plots that a typical half-power, or −3 dB, beam width is 30° rather than the desired 5°.

There are two ways in which a narrow beam is generally achieved, while still using such speakers. One is to re-shape the beam pattern by using a parabolic reflector, in much the same way as car headlights re-shape the broad beam from a light bulb into a narrow forward beam. The other method is to use multiple speakers, driven synchronously. The sound waves from multiple speakers reinforce in one direction and gradually cancel at angles further away from this direction. This is the principle of the phased array antenna.

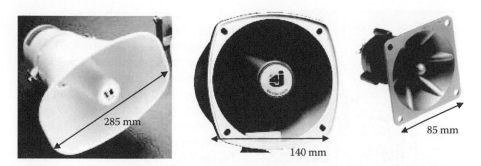

FIGURE 4.2 TOA SC630 (left), FourJay 440-8 (center), and Motorola KSN1005A (right).

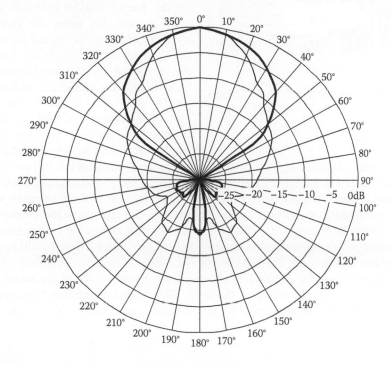

FIGURE 4.3 Polar response of some typical horn speakers, normalized to 0 dB in the forward direction. Heavy line (FourJay 440-8 at 3 kHz) and light line (TOA SC630 at 2 kHz).

4.2.2 DISH ANTENNAS

A parabolic dish antenna consists of a speaker situated at the focus of a parabolic reflector and facing downward toward the center of the reflector. An example is shown in Figure 4.4 and the geometry is shown in Figure 4.5.

It can be shown that if sound from the speaker is projected downward at an angle θ to the vertical, then its angle to the perpendicular from the dish surface is $\zeta = \theta/2$. The law of reflection gives $\zeta = \theta/2$ between the perpendicular and the verti-

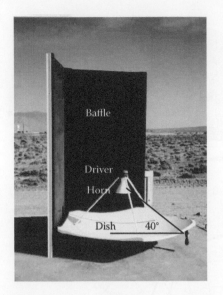

FIGURE 4.4 A SODAR dish antenna used in an early AV2000 AeroVironment SODAR.

cal. This means that all sound from the focal point is reflected directly upward and, regardless of the speaker's polar response, the upward beam is perfectly collimated. There are a number of reasons why this "perfect" situation is not observed in practice. The first is that the speaker cone has finite diameter d. The effect of this can be estimated using Figure 4.6.

In this figure we know that a downward ray from the center of the speaker (at the focus) will be reflected vertically upward. So, using the fact that sound propagation is reversible, a vertically downward ray from the edge of the speaker (at $x = d/2$) will be reflected back through the focus, and on past the speaker at an angle ψ to the vertical. If the sound intensity is uniform across the dish, then the beam will now have a width of approximately $\pm \psi = \pm d/2b$ rad.

For example, a dish having $b = 570$ mm and a speaker of diameter $d = 100$ mm would produce a beam nominally of width $\pm 5°$. In practice, the actual half-angle width of the beam (measured out to where the sound intensity is at half the intensity at the center of the beam) will depend on the angular or polar response of the speaker, and will generally be *less* than $\pm d / 2a$, where a is the dish radius. Very approximately, the speaker polar response within angles $-\tan^{-1}(b / a)$ to $\tan^{-1}(b / a)$, will be compressed into angles $-\tan^{-1}(d / 2b)$ to $\tan^{-1}(d / 2b)$.

A second cause for non-perfect collimation is whether some of the sound from the speaker reaches the edges of the dish. This creates diffraction (discussed in Chapter 3) with the upward traveling sound being equivalent to coming through a

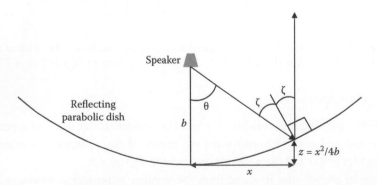

FIGURE 4.5 Geometry for a SODAR using a dish antenna. The downward-facing speaker is at the focal point of the parabolic dish.

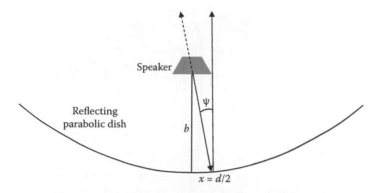

FIGURE 4.6 The effect of finite speaker diameter on dish antenna beam width.

hole with the same diameter as the dish. If the dish is uniformly covered by sound energy from the speaker, the upward intensity pattern is proportional to

$$\left[\frac{2J_1(ka\sin\psi)}{ka\sin\psi}\right]^2$$

where J_1 is the Bessel function of order 1, k the wavenumber of the sound, and a the dish radius. This gives a beam pattern which has an angular half-power width of about $\pm 2/ka$ rad, but which also has subsidiary peaks at greater angles (known as *side lobes*), as shown in Figure 4.7. For example, if $ka = 33$, then Figure 4.8 shows that the polar response of the diffraction pattern from the dish has a side lobe peak about 17 dB below the main lobe intensity and at an angle to the vertical of about $\psi = 9°$.

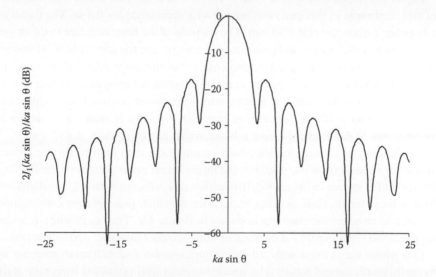

FIGURE 4.7 The polar intensity pattern from a uniformly radiated dish of radius a.

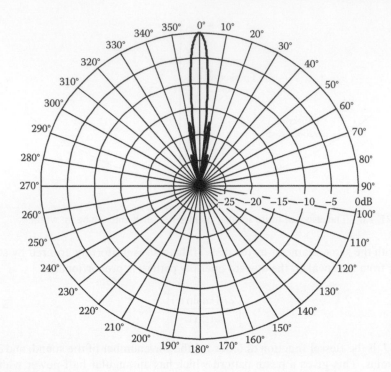

FIGURE 4.8   The polar pattern from a uniformly radiated dish of radius $a = 1.2$ m at a frequency of $f = 3$ kHz.

Again, the off-axis intensity will generally be lower than this because the sound power from the speaker will be concentrated more in the center of the dish.

Figure 4.4 shows a dish of $a \approx 0.6$ m radius and a speaker driver (the magnetic coil and diaphragm in this case) and horn at a focal distance $b \approx 0.4$ m. The throat of the horn has a diameter of $d \approx 70$ mm. The purpose of the horn attached to the driver is generally to efficiently couple the acoustic energy into the atmosphere. However, in this case the horn is designed to also ensure that the outer edges of the dish are subjected to minimal acoustic power so that diffraction is negligible. In other words, the driver/horn combination has a directivity with power confined well within the angle $\sin^{-1}(b / a)$ ($\approx 40°$ in this case, as shown in Fig. 4.4). If such a horn design is achieved, this SODAR should have a beam width narrower than $\pm d / 2b = \pm 5°$. The acoustically absorbing baffles which surround the dish help to further reduce sensitivity to sound from the side. For the rather broad polar pattern of the FourJay 440-8 at 3 kHz shown in Figure 4.3, diffraction is significant with a 1.2-m diameter, 580-mm focal length dish. A polar pattern for this dish plus speaker combination measured in an anechoic chamber is shown in Figure 4.9. The beam width is wider than predicted from Figure 4.8 because of finite speaker diameter and diffraction.

One advantage of the downward-facing horn speaker and dish arrangement is its inherently weatherproof nature. The speaker is quite well protected from rain. Rain noise, due to splashing on the dish, will still in general be a problem.

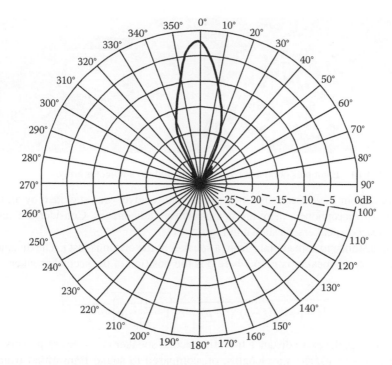

FIGURE 4.9 Measured polar pattern for the FourJay 440-8 speaker at 3 kHz with a 1.2-m diameter dish having a focal length of 580 mm.

Two manufacturers, AQS and Atmospheric Research, market small dish-antenna SODARs. Both argue that the antenna gives smaller side lobes than the alternative phased array. As we shall see later, smaller side lobes are desirable to reduce echoes from solid objects (such as masts, tress, or buildings). These systems use three speakers: each with its own dish. By mounting a speaker at the focal distance b but distance $x = -s$ to one side, the beam is tilted at angle $\tan^{-1}(s/b)$ rad in the $+x$-direction. This provides the three measurements of Doppler shift needed at each height to find the three unknown wind velocity components u, v, and w.

4.2.3 PHASED ARRAY ANTENNAS

Most SODAR designs use multiple speakers in a phased array. There are two basic types: a horizontal array (with speakers facing upward) and a reflector-array (with speakers facing approximately horizontally toward a 45° reflector). In the first case, speakers must be protected from rain by being a folded or re-entrant horn design (Figure 4.10). In the second case, any speaker may be used, and the array is recessed into a rain shield. This design is perhaps a little less susceptible to rain impact noise, but is generally bulkier. The beam geometry is essentially the same for the two cases, however.

FIGURE 4.10   Examples of (a) a horizontal array and (b) a reflector array.

Apart from the extra transmitted power and receiver area provided by an array of speaker/microphones, there are beam-forming advantages. Consider the case of evenly spaced speakers, as shown in Figure 4.11.

The distance to some point $\underline{r}$ from a speaker at the origin is just r. The distance to r from a second speaker is $\underline{r} - \underline{\rho}$ where $\underline{\rho}$ is the position of the second speaker. Now

$$|\underline{r} - \underline{\rho}| = \sqrt{(\underline{r} - \underline{\rho}) \cdot (\underline{r} - \underline{\rho})} = r \sqrt{1 - \frac{2}{r}\left(\frac{\underline{r}}{r} \cdot \underline{\rho}\right) + \frac{\rho^2}{r^2}}.$$

For $r \gg \rho$, the *extra* distance from the second speaker is $\approx -(\underline{r}/r) \cdot \underline{\rho}$. This extra distance is $-(\underline{r} \cdot \underline{\rho})/r\lambda$ wavelengths, or, compared to sound transmitted from the origin, a phase angle of $-k(\underline{r}/r) \cdot \underline{\rho}$, where k is the wavenumber. If the position of the second speaker is (x, y), then the phase of sound from this speaker, compared to sound transmitted from the origin, is

$$\gamma = -k(x\cos\phi\sin\theta + y\sin\phi\sin\theta) + x\frac{\partial\varphi}{\partial x} + y\frac{\partial\varphi}{\partial y}. \tag{4.1}$$

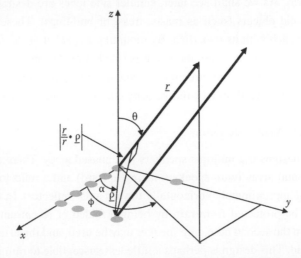

FIGURE 4.11   The geometry of an array of speakers, some of which are shown as gray dots, transmitting in a direction $\underline{r}$.

Allowance has been made in (4.1) for an extra phase shift $\varphi(x, y)$ to be applied electronically to the sound transmitted from a speaker at (x, y). It is also assumed that φ increases or decreases by uniform steps from speaker to speaker across the array. We shall see below that this progressive change of applied phase across the array allows the acoustic beam to be "steered" in space.

Let $(x, y) = (md, nd)$, where d is the inter-speaker spacing in both x and y directions, and write

$$\gamma = m\left(-kd\cos\phi\sin\theta + d\frac{\partial\varphi}{\partial x}\right) + n\left(-kd\sin\phi\sin\theta + d\frac{\partial\varphi}{\partial y}\right) = m\varphi_x + n\varphi_y.$$

Assume that the signal transmitted from the speaker at the origin is $a(\theta,\phi)\sin\omega t$. The direction-dependent amplitude $a(\theta,\phi)$ allows for the speaker polar pattern.

Assume also, for the moment, that $\partial\varphi/\partial y = 0$. Then in the plane $\phi = 0$, the incremental phase $\varphi_y = 0$. The contribution to the sound signal at a distant point from the two rows $-m$ and $+m$ in this plane is therefore

$$a(\theta,0) \sum_{n=-(N_m-1)/2}^{(N_m-1)/2} [\sin(\omega t - m\varphi_x) + \sin(\omega t + m\varphi_x)] = [2a(\theta,\phi)N_m\cos(m\varphi_x)]\sin\omega t,$$

assuming there are N_m speakers in these rows. If there are a total of M rows, then the total amplitude is

$$A(\theta,0) = a(\theta,0)N_0 + 2a(\theta,0) \sum_{m=1}^{(M-1)/2} N_m\cos(m\varphi_x).$$

Assume that the speaker polar response $a(\theta,\phi)$ changes only slowly with angle and so

$$\frac{\partial a(\theta,0)}{\partial\theta} \approx 0.$$

The amplitude therefore peaks when

$$\frac{\partial A}{\partial\theta} = 2a(\theta,0)kd\cos\theta \sum_{m=1}^{(M-1)/2} N_m m\sin(m\varphi_x) = 0.$$

This is true when

$$\varphi_x = -kd\sin\theta + d\frac{\partial\varphi}{\partial x} = 0, \pm 2\pi, \pm 4\pi, \ldots \qquad (4.2)$$

or

$$kd\sin\theta = d\frac{\partial\varphi}{\partial x}, d\frac{\partial\varphi}{\partial x} \pm 2\pi, \ldots. \qquad (4.3)$$

For example, if the incremental applied phase step is $d(\partial\varphi/\partial x)=\pi/2$ rad, then peaks occur at

$$\sin\theta=\ldots,-\frac{3\lambda}{4d},\frac{\lambda}{4d},\frac{5\lambda}{4d},\ldots. \tag{4.4}$$

The choice of a speaker–speaker phase increment of $\pi/2$ is an important one, since it is very easy to electronically generate signals $\sin\omega t$, $\cos\omega t=\sin(\omega t+\pi/2)$, $-\sin\omega t=\sin(\omega t+2\pi/2)$, and $-\cos\omega t=\sin(\omega t+3\pi/2)$. This phasing is shown in Figure 4.12.

The above analysis shows that it is relatively easy to tilt a phased array beam electronically. This beam steering is useful for obtaining Doppler shift from wind components projected onto the beam direction.

Figure 4.13 shows an example of angle for peak intensity versus frequency for one speaker type. Since the maximum SPL from a speaker generally occurs at a wavelength related to the speaker diameter, the optimum frequency of operation for the phased array, if using 90° phase steps, generally gives a tilt angle in the range 15–25°. Note, however, that a common feature of phased array beam steering is the appearance of multiple peaks, as predicted by (4.3). In Figure 4.13, three peaks occur at high frequencies. At 6 kHz, peaks occur at −30°, 9.6°, and 56°. The natural speaker response $a(\theta,\phi)$ will tend to suppress the 56° peak, but the −30° peak could be troublesome as a source for spurious echoes off fixed objects such as trees, buildings, and masts. Such signals are called "fixed echoes" and are a significant design limitation of many SODARs. Because of this appearance of multiple lobes, it is common to phase the array to tilt the beam *diagonally*, thus giving a speaker row spacing of $d/\sqrt{2}$. As shown in Figure 4.13, this causes the side lobes to be at lower elevation angles and therefore to be more suppressed by the speakers directional response $a(\theta,\phi)$.

The above analysis assumes that there is a central speaker and symmetrically placed speakers on either side. If instead $(x,y)=(md-d/2,nd-d/2)$ then

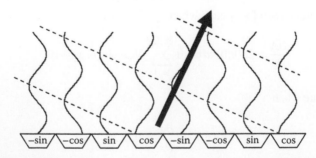

FIGURE 4.12 Snapshot of pressure waves transmitted from a row of speakers with incrementally increasing phase of $\pi/2$ to the right. Dashed lines show wavefronts and the solid arrow shows the propagation direction.

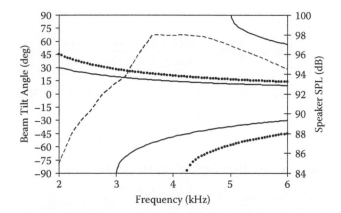

FIGURE 4.13 The angle for peak intensity when 90° phase steps are used with KSN1005A speakers as a function of frequency. Solid lines: row spacing d; dots: row spacing $d/\sqrt{2}$. Also shown is the SPL versus frequency for these speakers (dashed line).

$$A(\theta,0) = 2a(\theta,0)\sum_{m=1}^{M/2} N_m \cos\left(\left[m-\frac{1}{2}\right]\varphi_x\right)$$

and (4.3) still holds.

Equation (4.3) gives the angular *position* of the beam maximum, and this is independent of the number of speakers (although there must be at least two speakers in a row). The number of speakers affects the *width* of the acoustic beam and also the nature of subsidiary maxima. Consider the case where there is no central speaker. Then following a similar analysis to that above, the total amplitude is

$$A(\theta,\phi) = 8a(\theta,\phi)\sum_{m=1}^{M/2}\cos\left(\left\{m-\frac{1}{2}\right\}\varphi_x\right)\sum_{n=1}^{N/2}\cos\left(\left\{n-\frac{1}{2}\right\}\varphi_y\right)$$

providing there are equal numbers N of speakers in each row. Now using the identities $\sin\theta = (e^{j\theta} - e^{-j\theta})/2j$ and $\cos\theta = (e^{j\theta} + e^{-j\theta})/2$ gives

$$\sum_{m=1}^{M/2}\cos\left(\left\{m-\frac{1}{2}\right\}\varphi_x\right) = \frac{1}{2}\sum_{m=1}^{M/2}e^{j(m-1/2)\varphi_x} + \frac{1}{2}\sum_{m=1}^{M/2}e^{-j(m-1/2)\varphi_x}$$

$$= \frac{1}{2}e^{-i(1/2)\varphi_x}\sum_{m=1}^{M/2}(e^{j\varphi_x})^m + \frac{1}{2}e^{j(1/2)\varphi_x}\sum_{m=1}^{M/2}(e^{-j\varphi_x})^m.$$

The sums can be evaluated as geometric series, giving

$$\sum_{m=1}^{M/2}\cos\left(\left\{m-\frac{1}{2}\right\}\varphi_x\right) = \frac{1}{4}\frac{\sin(M/2)\varphi_x}{\sin(\varphi_x/2)},$$

$$A(\theta,\phi) = \frac{1}{2}a(\theta,\phi)\frac{\sin(M/2)\varphi_x}{\sin(\varphi_x/2)}\frac{\sin(N/2)\varphi_y}{\sin(\varphi_y/2)}. \tag{4.5}$$

Aside from the main peaks at $\varphi_x = \varphi_y = 0, \pm\pi, \pm 2\pi, \ldots$, there are $M-1$ nulls at

$$\frac{M}{2}\varphi_x = \pm\pi, \pm 2\pi, \ldots, \pm(M-1)\pi$$

for any particular φ_y and therefore subsidiary peaks at the $M-2$ intermediate positions

$$\varphi_x \approx \pm\frac{3\pi}{M}, \pm\frac{5\pi}{M}, \ldots, \pm\frac{(2M-3)\pi}{M}. \tag{4.6}$$

There are therefore *side lobes* having relative *intensities* of

$$\frac{I_{mn}}{I_0} = \left[\left\{\frac{1}{M\sin((2m+1)\pi/2M)}\right\}\left\{\frac{1}{N\sin((2n+1)\pi/2N)}\right\}\right]^2,$$

$$m = 1, 2, \ldots, M-2, n = 1, 2, \ldots, N-2 \tag{4.7}$$

compared to the main lobe, plus the fall off with increasing angle due to the individual speaker response $a(\theta,\phi)$. The width of the main lobe to the first null is

$$\Delta\theta \approx \frac{2\pi}{Mkd}. \tag{4.8}$$

Figures 4.14 and 4.15 show beam patterns for $kd = 5$. The second "main" lobe problem is very evident in Figure 4.15.

4.2.4 Antenna Shading

The beam patterns derived above for phased arrays are multiplied by the individual speaker response $a(\theta,\phi)$ as shown, for example, by (4.5). For a square array, the basic

$$\frac{\sin(M/2)\varphi_x}{\sin(\varphi_x/2)}$$

dependence is really just the square aperture version of the circular antenna diffraction dependence discussed earlier in this chapter for dish antennas. Using the array symbols, this would be

$$\frac{2J_1((M/2)kd\sin\theta)}{\sin((1/2)kd\sin\theta)},$$

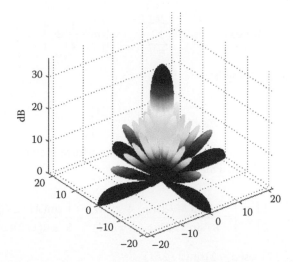

FIGURE 4.14 (See color insert following page 10). The beam pattern from an 8 × 8 square array without an applied phase gradient and with $kd = 5$.

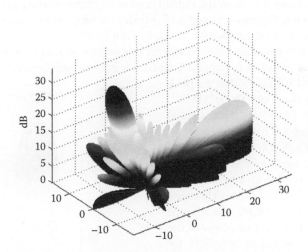

FIGURE 4.15 (See color insert following page 10). The beam pattern from an 8 × 8 square array with an applied phase increment of $\pi/2$ per speaker and with $kd = 5$.

which closely resembles the square array beam shape. The array pattern can therefore be thought of as being due to the output of an infinite array of speakers limited by a square hole with a resulting diffraction pattern.

Now if a square aperture were radiated by sound which had intensity *reduced* near the edges of the hole, then the diffraction effect would also be reduced. This leads to the idea of antenna "shading" in which the gain of speakers is reduced toward the outside of the array. Then

$$A(\theta,0) = 2a(\theta,0)N \sum_{m=1}^{M/2} w_m \cos\left(\left[m - \frac{1}{2}\right]\varphi_x\right).$$

One common weighting is

$$w_m = \cos\left(\frac{\pi[m - 1/2]}{M}\right). \tag{4.9}$$

The response A can be found using Fourier transform methods, but also using the method leading to (4.5). This gives

$$\sum_{m=1}^{M/2} w_m \cos\left(\left\{m - \frac{1}{2}\right\}\varphi_x\right) = \frac{1}{8}\frac{\sin(M/2)(\varphi_x + \pi/M)}{\sin(1/2)(\varphi_x + \pi/M)} + \frac{1}{8}\frac{\sin(M/2)(\varphi_x - \pi/M)}{\sin(1/2)(\varphi_x - \pi/M)}$$

$$= \frac{-\cos(M\varphi_x/2)\cos(\varphi_x/2)\sin(\pi/2M)}{4\sin(\varphi_x/2 + \pi/2M)\sin(\varphi_x/2 - \pi/2M)}. \tag{4.10}$$

Figures 4.16 and 4.17 show the shaded responses corresponding to the unshaded responses shown in Figures 4.14 and 4.15. Minor side lobes are strongly suppressed, but the shading does not remove the multiple main beams.

There are two penalties associated with this improved side lobe structure. The first is that less power is transmitted, since the gain of speakers is reduced. Putting $\varphi_x = 0$ in (4.10) shows that the peak intensity value is about $(16/\pi^4)M^2N^2$ compared with M^2N^2 for the unshaded case. This is about 8 dB loss in peak intensity for $M = N = 8$. The second penalty is that the main lobe is wider. The first null now occurs at $\theta \approx 3\pi/Mkd$ instead of at $\theta \approx 2\pi/Mkd$. However, as shown in Figure 4.18, the shaded array has nearly 80% of its power in the main beam, compared to only 50% for the unshaded array.

Shading can be accomplished via

1. a passive attenuator at each speaker,
2. feeding signals of differing amplitude individually to each speaker, or
3. applying incremental phase shifts of π/M and $-\pi/M$ in the x-direction and π/N and $-\pi/N$ in the y-direction.

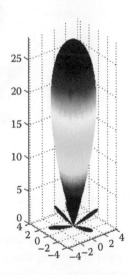

FIGURE 4.16 (See color insert following page 10). The beam pattern from an 8×8 square array without an applied phase gradient and with $kd = 5$ and a cosine-shaded speaker gain pattern.

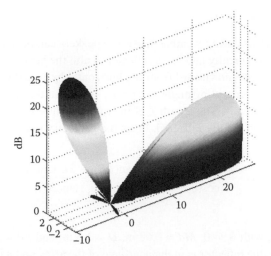

FIGURE 4.17 (See color insert following page 10). The beam pattern from an 8 × 8 square array with an applied phase increment of $\pi/2$ per speaker and with $kd = 5$ and a cosine-shaded speaker gain pattern.

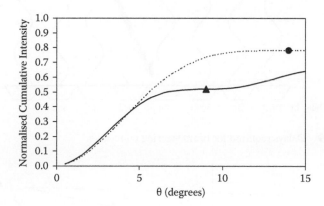

FIGURE 4.18 The normalized cumulative intensity outward from the vertical for an 8 × 8 square array with no applied phase increment, and with $kd = 5$. Solid line: unshaded; dashed line: cosine shaded; triangle: position of null for unshaded array; circle: position of null for shaded array.

Both the second and third methods require separate signals to each speaker.

4.2.5 RECEIVE PHASING

Beam steering for reception of an echo signal with a phased array requires a progressive phase shift in the *opposite sense* to that used for transmission. So, for example, increasing the phase to each speaker in the +x-direction by $\pi/2$ during transmission requires *delaying* successively in the +x-direction by $\pi/2$, as shown in Figure 4.19.

4.2.6 REFLECTORS

Phased array SODARs using weather-sensitive speakers can have the speaker array mounted facing horizontally and use a reflector to aim the beam vertically.

If the array is tilted downward from the horizontal by angle α then the reflector must be tilted from vertical by angle $\theta_r = \theta_s / 2 + \pi / 4$ so that the un-phased main lobe is directed vertically. The length Z of the reflector must be sufficiently large so that the phased beam is fully reflected. From the geometry in Figure 4.20,

$$Z = \frac{Md[1 + \tan(\theta_r - \theta_s)\tan(\Delta\theta)] + 2D\tan(\Delta\theta)}{[\cos(\theta_r - \theta_s) + \sin(\theta_r - \theta_s)\tan(\Delta\theta)]}.$$

For example, with $\theta_s = 0$, $Md = 0.68$ m, $D = 1.5$ m, and $\Delta\theta = 20°$, $Z = 1.42$ m. Strictly speaking, the reflector is in the near-field of the array and a little more length

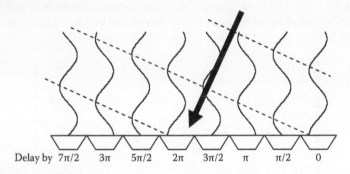

Delay by $7\pi/2$ $\quad 3\pi$ $\quad 5\pi/2$ $\quad 2\pi$ $\quad 3\pi/2$ $\quad \pi$ $\quad \pi/2$ $\quad 0$

FIGURE 4.19 Delays required for beam steering on receive.

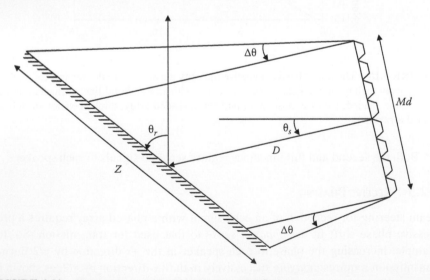

FIGURE 4.20 The geometry of a phased array and reflector.

should be allowed. Any reflected sound must also clear the array and weather shield-ing around the array: this determines the length D.

The reflector must be a good acoustic reflector but not be resonant or be noisy when rain splashes on it. Reflectors are generally constructed from marine plywood or from fiberglass and, since sound might penetrate the reflector and cause problems with spurious fixed echoes, the reflector will normally be backed with acoustic paint, lead, and/or acoustic foam.

4.3   MONOSTATIC AND BISTATIC SODAR SYSTEMS

All commercial SODARs are *monostatic*, which means that the transmitter antenna and the receiver antenna are in the same position. This arrangement has advantages of compactness, simple geometry and interpretation, simpler deploy-ment, and generally lower cost due to use of the same transducers as speakers and microphones.

Bistatic configurations use spatially separated transmitters and receivers. This is common for the microwave transmitter and receiver on a radio acoustic sounding system. The general geometry is shown in Figure 4.21.

The scattering geometry is determined by the baseline distance D, the transmit-ter and receiver tilt angles θ_T and θ_R, and the orientation of the transmit plane φ_T and of the receive plane φ_R with respect to the baseline. From the geometry shown in Figure 4.21,

$$z = \frac{D}{\tan\theta_R \cos\phi_R + \tan\theta_T \cos\phi_T},$$

$$r_T = \frac{z}{\cos\theta_T},$$

$$r_R = \frac{z}{\cos\theta_R},$$

$$\cos\beta = \cos\theta R\left(\frac{D}{z}\sin\theta_T \cos\phi_R - \frac{1}{\cos\theta_T}\right).$$

(4.11)

The *scattering angle* is β. For the monostatic case, $\theta_T = \theta_R$ and $D=0$, giving cos $\beta = -1$ or $\beta = \pi$.

One of the difficulties with a bistatic system is aiming the beams so that adequate intersection occurs. Generally it is simpler to have the transmitter beam vertical and several receiver beams tilted. Then $\beta = \pi - \theta_R$ and $z = D/\tan\theta_R$.

For monostatic SODARs, the vertical profiling is achieved by taking the time record of the echo as being a distance record. For bistatic SODARs, the range is set by the intersection of the transmitted and received main lobes. So it is necessary to scan the tilted receiver beam using phasing. This is best done by mechanically tilt-ing the receiver so that its un-phased beam points to the middle of the height range

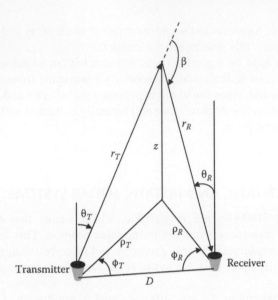

FIGURE 4.21   The geometry of bistatic systems.

of interest. Then the effect of secondary main lobes appearing is minimized. This
geometry is shown in Figure 4.22.

In this case

$$\theta_R = \tan^{-1}\frac{z}{D + z\tan\theta_T} - \tan^{-1}\frac{z_0}{D} \qquad (4.12)$$

for the 2D situation shown. Using the same parameters as in Figures 14–17, the
product of the transmitted and received intensities gives a measure of system
sensitivity. This quantity is plotted in Color Figure 4.23 for the unshaded case,
and in Color Figure 4.24 for the shaded case, with $z_0 = 50$ and 100 m. There are
two very significant features of these plots. The first is that the side lobes in the
unshaded case will give echoes from a wide range of locations, both horizontally
and vertically. Since there is no guarantee of any degree of homogeneity in the
turbulence intensity which determines the echo strength, this can mean that the
echoes come from an unexpected height or even from an unexpected angle (which
is important for wind measurements, as will be seen later). The second feature
to note is that, even in the shaded case, the height resolution is very coarse as
z_0 increases. The net result is that, for bistatic systems to work effectively, it is
essential that the beams be well defined and have minimal side lobes and also
that the systems be pulsed rather than continuous. For example, a bistatic system
having a pulse which is 10-m long would have the positional sensitivity shown in
Figure 4.25.

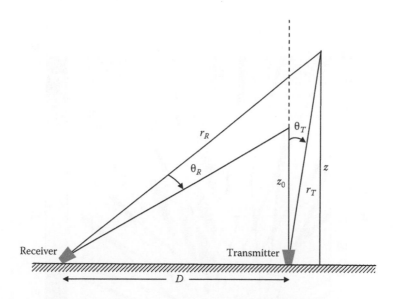

FIGURE 4.22 The geometry for a simplified bistatic configuration.

4.4 DOPPLER SHIFT FROM MONOSTATIC AND BISTATIC SODARS

In the presence of air flow, the frequency of the echo signal changes (i.e., is Doppler shifted), allowing the wind speed to be estimated. This is the most used feature of SODARs. There are many textbook derivations of Doppler shift, but it is very nearly impossible to find a general treatment of reflection from a target moving with the medium. Treatments which have appeared in journal papers are generally incorrect. A treatment for the bistatic case (Georges and Clifford, 1972) was then extended with examples for the monostatic situation (Georges and Clifford, 1974). Unfortunately their formula for Doppler shift does not reduce to the simple 1D textbook case when the transmitter, receiver, and wind are in line. This also means that numerical simulations based on the Georges and Clifford formulae by Phillips et al. (1977) and Schomburg and Englich (1998) are suspect. More recently, Ostashev (1997) has treated the 2D (x, z) case and has found that the error in wind speed is $\Delta V = (V^2 / c)\sin\theta$. This is in contrast to Georges and Clifford, who find that there is no refractive correction (to second order) when the wind is entirely horizontal. Given the confusion in these various treatments, and the need for a 3D correction formula, we now give a basic derivation of "beam drift" effects.

We will explain what happens through a simple description of the time taken for two successive wavefronts to travel from the transmitter via reflection off turbulence to the receiver. The time difference between the arrival times of the two wavefronts at the receiver gives the period of the Doppler-shifted echo signal. Figure 4.26 shows this concept with a tilted transmitter and a vertical receiver. Although the transmitted beam is aimed to intersect directly above the receiver, it is blown downwind during its upward journey. Similarly, it is not the reflected sound aimed directly at the receiver which reaches it, but rather sound which is initially directed somewhat

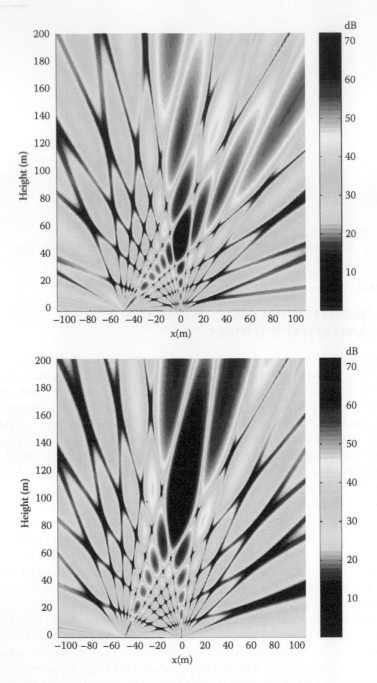

FIGURE 4.23   (See color insert following page 10). Unshaded bistatic system sensitivity for baseline $D = 50$ m, and with preset intersection height $z_0 = 50$ m and 100 m.

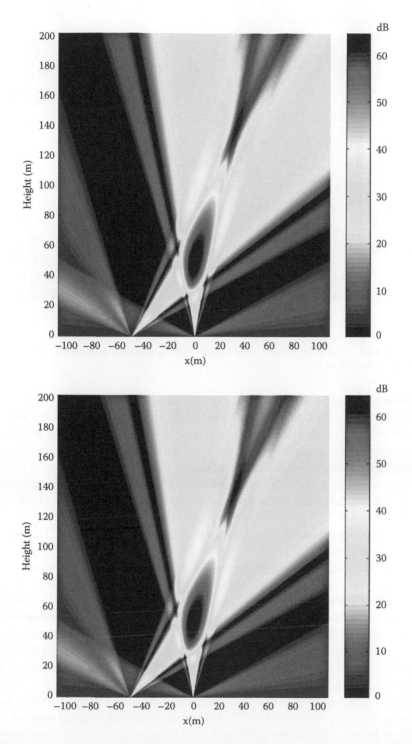

FIGURE 4.24 (See color insert following page 10). Shaded bistatic system sensitivity for baseline $D = 50$ m, and with preset intersection height $z_0 = 50$ m and 100 m.

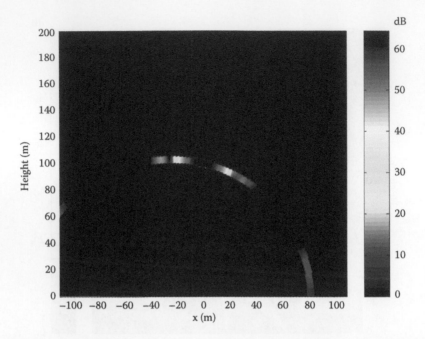

FIGURE 4.25 Positional sensitivity of a gated bistatic shaded phased array system having a pulse length of 10 m.

upstream of the receiver. The net result is that the "round trip" distance and time are different from the no-wind case. By the time a second wavefront leaves the transmitter, the turbulent patch will have moved further downstream and so the path to it is different from that of the first wavefront. Because of these changing paths, the time between arrivals of the two wavefronts at the receiver is in general different from the time between departures of the wavefronts from the transmitter. This means that the period and the frequency of the detected signal are different from the transmitted signal.

To illustrate this, consider the tilted beam in a monostatic SODAR. The situation is shown in Figure 4.27.

Sound is transmitted upward at an angle θ to the vertical, aimed toward a turbulent patch. This sound takes a time t_s to reach height z. During this time, the patch of turbulence has moved horizontally a distance Vt_s as shown, where V is the wind speed. When the sound meets the turbulence at height z, but distance Vt_s downstream from the original position of the turbulence, sound is scattered in all directions. Some of this scattered sound, initially aimed downward but upstream of the SODAR, will reach the SODAR after a further time t_r, as shown in the left-hand plot of Figure 4.27 (c is the sound speed, assumed to be uniform in this example).

Sound emitted one period T later is also initially aimed at the turbulent patch, which at this time is a distance VT downstream from the original position. The sound paths are shown in the right-hand plot of Figure 4.27, in which the times for upward and downward propagation are slightly different from the left-hand diagram,

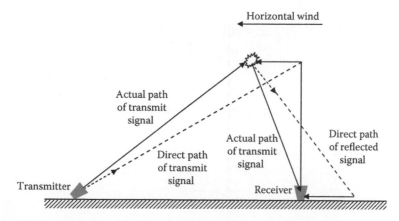

FIGURE 4.26 The basic concept of extra path when there is a wind.

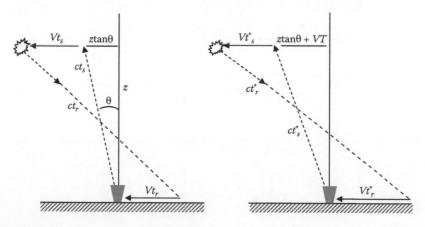

FIGURE 4.27 Sound path in the simple monostatic, tilted beam case, at two times separated by one period of the transmitted sound.

as indicated by the primes on the times. The geometry is a little clearer if the triangles are redrawn as in Figure 4.28.

In these figures, the horizontal movement due to the wind has been grossly exaggerated, since $V \ll c$. In this case, $\sin \Delta\theta \approx \Delta\theta$ and $\cos \Delta\theta \approx 1$, so

$$ct_r = ct_s \cos\Delta\theta + V(t_r + t_s)\sin(\theta + \Delta\theta) \approx ct_s + V(t_r + t_s)\sin\Delta\theta$$

giving

$$ct_r\left(1 - \frac{V}{c}\sin\theta\right) \approx ct_s\left(1 + \frac{V}{c}\sin\theta\right)$$

or

$$ct_r \approx ct_s \left(1 + 2\frac{V}{c}\sin\theta\right).$$

Similarly

$$ct_r^* \approx ct_s^* \left(1 + 2\frac{V}{c}\sin\theta^*\right),$$

but

$$ct_s = \frac{z}{\cos\theta}$$

and

$$ct_s^* = \frac{z}{\cos\theta^*}.$$

Also

$$\tan\theta^* = \frac{z\tan\theta + VT}{z} = \tan\theta + \frac{VT}{z},$$

so

$$\frac{1}{\cos\theta^*} = \left[1 + \left(\tan\theta + \frac{VT}{z}\right)^2\right]^{1/2} \approx \frac{1}{\cos\theta}\left[1 + \left(\tan\theta\frac{VT}{z} + \frac{V^2 T^2}{2z^2}\right)\cos^2\theta\right].$$

Finally, the time between "round trips" for two parts of the signal transmitted one cycle apart is the period T^* of the echo signal at the receiver

$$T^* = (t_r^* + t_s^*) - (t_r + t_s) = 2t_s^*\left(1 + \frac{V}{c}\sin\theta^*\right) - 2t_s\left(1 + \frac{V}{c}\sin\theta\right) \approx \left(2\frac{V}{c}\sin\theta + 2\frac{V^2}{c^2}\right)T.$$

The shift in frequency is

$$\frac{\Delta f}{f_T} = \left(\frac{1}{T^*} - \frac{1}{T}\right)T = -2\frac{V}{c}\sin\theta - 2\frac{V^2}{c^2}. \tag{4.13}$$

The first term on the right, $-2(V/c)\sin\theta$, is the usual Doppler shift term used to calculate the wind speed component V from the measured shift Δf in the position of the peak in the frequency spectrum, given the known beam tilt angle θ. For typical SODAR systems, $\theta = 15°$ to $25°$, so the last term on the right is never greater than about a third of the magnitude of the first term on the right.

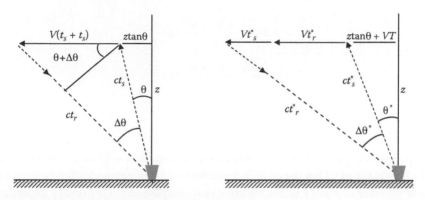

FIGURE 4.28 Redrawn geometry from Figure 4.27.

For example, with $\theta = 18°$, and $V = 14$ m s^{-1}, $\Delta f / f_T = -0.0216$ and the estimated velocity would be $\hat{V} = -(c/2\sin\theta)(\Delta f / f_T) = 12$ m s^{-1}, a 20% error. With $V = -14$ m s^{-1}, a -20% error occurs. Note that this *beam drift effect* gives a bias in derived winds at higher wind speeds, causing the estimated wind speed to be lower if the wind is away from the SODAR and to be higher if the wind is towards the SODAR. The direction of reception also changes a little due to the second-order V^2 term. For example, if $\theta = 18°$, $u = 2$ m s^{-1}, and $v = 10$ m s^{-1}, $V = 10.2$ m s^{-1}, wind direction $\psi = 11°$ and $\Delta\psi = 4°$.

If a 3-beam monostatic SODAR is aligned with its beams in the u and v directions

$$\eta_1 = \hat{u}\sin\theta + \hat{w}\cos\theta = u\sin\theta + w\cos\theta - \frac{u^2 + v^2 + w^2}{c},$$

$$\eta_2 = \hat{v}\sin\theta + \hat{w}\cos\theta = v\sin\theta + w\cos\theta - \frac{u^2 + v^2 + w^2}{c},$$

$$\eta_3 = \hat{w} = w - \frac{u^2 + v^2 + w^2}{c},$$

from which

$$\hat{u} = \frac{\eta_1 - \eta_3\cos\theta}{\sin\theta},$$

$$\hat{v} = \frac{\eta_2 - \eta_3\cos\theta}{\sin\theta},$$

$$\hat{w} = \eta_3.$$

The set of equations can be solved for u, v, and w, but a simpler approximate solution is found by putting $u^2 + v^2 + w^2 = \hat{u}^2 + \hat{v}^2 + \hat{w}^2$ in the correction term. Then

$$u \approx \hat{u} + \frac{\hat{u}^2 + \hat{v}^2 + \hat{w}^2}{c} \tan\frac{\theta}{2},$$

$$v \approx \hat{v} + \frac{\hat{u}^2 + \hat{v}^2 + \hat{w}^2}{c} \tan\frac{\theta}{2},$$

$$w \approx \hat{w} + \frac{\hat{u}^2 + \hat{v}^2 + \hat{w}^2}{c}.$$

For a pointing angle of $\theta = 18°$ and a total wind speed $V = 15$ m s^{-1}, the error in u and v is only 0.1 m s^{-1} but the error in w is 0.66 m s^{-1}. The reason for the large reduction in the error in u and v is that the error in w compensates. This is very important, since it means that the vertical velocity *must* be included in the estimation of u and v. The total wind speed is

$$V \approx \hat{V}\left[1 + \frac{(\hat{u} + \hat{v})\tan(\theta/2) + \hat{w}}{c}\right].$$ (4.14)

The error in ignoring the beam drift is therefore both speed- and direction-dependent. Depending on the range of wind directions encompassed, curvature of a calibration may not be evident (i.e., if $u \approx -v$). However, the largest error, when say $v = w = 0$, is only around 1% at 20 m s^{-1}. This means, providing the vertical velocities are included properly, that beam drift errors are unlikely to be evident.

For general vector wind $\underline{V}$ and transmitter–receiver configurations, this is a 3D problem and both the diagrams and the mathematics are a bit more complex. Although not proven here, in the general case the relative shift in frequency is

$$\frac{\Delta f}{f_T} \approx -(\hat{\underline{r}}_R + \hat{\underline{r}}_T)\cdot\frac{V}{c} - \frac{V^2}{c^2}\left(1 + \frac{R_T}{R_R}\right) - \left(\hat{\underline{r}}_R \cdot \frac{V}{c}\right)\left(\hat{\underline{r}}_T - \frac{R_T}{R_R}\hat{\underline{r}}_R\right)\cdot\frac{V}{c},$$ (4.15)

where $\hat{\underline{r}}_T$ is a unit vector in direction pointed to by the transmitter and $\hat{\underline{r}}_R$ is a unit vector in direction pointed to by the receiver. The distance from the transmitter to the turbulent patch is R_T and the distance from the receiver to the turbulent patch is R_R. Once again there is a wind-direction-dependent term and smaller terms dependent on V^2.

For the case where the wind has components (u, v, w), (4.15) predicts for the monostatic case

$$\frac{\Delta f}{f_T} \approx -\frac{2}{c}(u\sin\theta\cos\phi + v\sin\theta\sin\phi + w\cos\theta)$$ (4.16)

for the relative frequency shift from a beam pointing in the (θ, θ) direction (with wind component u in the horizontal plane $\theta = 0$, and *toward* direction $\phi = 0$), and ignoring the second-order term.

For the bistatic case shown in Figure 4.29,

$$\frac{\Delta f}{f_T} \approx \frac{D}{R_R}\left(\frac{u}{c}\cos\phi + \frac{v}{c}\sin\phi\right) - \frac{w}{c}\left(1 + \frac{z}{R_R}\right) - 2\frac{V^2}{c^2}.$$

In the case of the tilted transmitted beam, shown in Figure 4.30,

$$\frac{\Delta f}{f_T} = -\left[z\tan\theta\left(\frac{1}{R_T} + \frac{1}{R_R}\right) - \frac{D}{R_R}\right]\frac{u}{c} - z\left(\frac{1}{R_T} + \frac{1}{R_R}\right)\frac{w}{c} \qquad (4.17)$$

with $R_T = z/\cos\theta$ and $R_R = [(D - z\tan\theta)^2 + z^2]^{1/2}$. The term in u is positive for $\tan\theta > D/2z$ and negative for larger θ. This means that for a vertically transmitting bistatic SODAR, low-elevation side lobes can give a forward-scattered signal which has a Doppler shift opposite in sign from the correct signal.

Such a combined spectrum, including the influence of side lobes, is shown from the Heimdall continuously transmitting bistatic SODAR (Mikkelsen et al., 2005). The narrow spectral peak at 3960 Hz is the direct signal from the transmitter to the receiver. The broader spectral peak to the left is due to the vertically transmitted pulse. Note that it is much broader than the direct signal spectrum because of the wide range of angles β for this continuous system. There is also a second broader peak, partly underlying the direct signal peak and slightly to its right. This is due to a diffraction side lobe from the dish antenna used. Given that $f_T = 3960$ Hz, and the peak at the left is at 3920 Hz (for $\theta = 0$), $u/c = (40/3960)R_R/D$. The half-width of the left-hand spectral peak is about 50 Hz, so the range of scattering angles, expressed as $\Delta\theta$, is $\Delta\theta = (50/3960)/[(u/c)(1+z/R_R)] = (50/3960)/[(40/3960)(R_R+z)/D] = 1.25[D/(R_R+z)]$. Here $D = z$, so $\Delta\theta = \pm30°$, which emphasizes the need for bistatic SODARs to be pulsed systems. The broad peak at the right, at 3970 Hz, will be from a side lobe at about 27° from the vertical. Side lobes at such angles readily exist since they will generally be within the angular pass region of acoustic baffles. For monostatic SODARs, such a side lobe would be unlikely to cause problems, but in the case of the bistatic system it is significant.

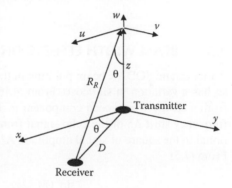

FIGURE 4.29 The geometry of a bistatic SODAR.

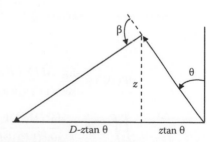

FIGURE 4.30 The geometry for a tilted transmitting beam bistatic SODAR.

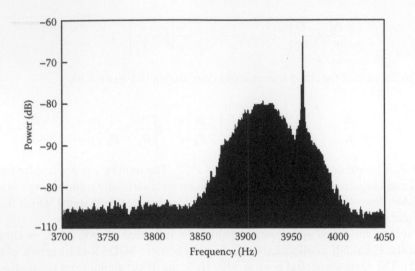

FIGURE 4.31   The spectrum recorded by the Heimdall bistatic SODAR.

4.5   BEAM WIDTH EFFECTS ON DOPPLER SHIFT

A monostatic SODAR beam pointing in the x–z-plane has a finite angular width and so has a variation in $\sin\theta\cos\phi$. From (4.16), a range of frequency shifts will result from a steady wind speed component u. The spectral power density, dP/df, at a frequency shift Δf, of the echo signal from a small volume element will be proportional to the square of signal amplitude $A^2(\theta,\phi)$, and the size of the volume element. From (4.5)

$$dP \propto a^2(\theta,\phi)\frac{\sin^2(M/2)\varphi_x}{\sin^2(\varphi_x/2)}\frac{\sin^2(N/2)\varphi_y}{\sin^2(\varphi_y/2)}\sin\theta\,d\theta\,d\phi.$$

But from (4.15), for the case of a beam tilted in the x-direction,

$$\varphi_x = d\frac{\pi\Delta f}{u} + d\frac{\partial\varphi}{\partial x},$$

so

$$dP(\Delta f) \propto \frac{\sin^2(M/2)(d(\pi\Delta f/u)+d(\partial\varphi/\partial x))}{\sin^2(1/2)(d(\pi\Delta f/u)+d(\partial\varphi/\partial x))}$$

$$\iint a^2(\theta,\phi)\frac{\sin^2(N/2)kd\sin\phi\sin\theta}{\sin^2(1/2)kd\sin\phi\sin\theta}\sin\theta\,d\theta\,d\phi,$$

where the integral includes all volume elements where $\sin\theta\cos\phi = -c\Delta f / 2uf_T$. In practice, the integral does not depend much on Δf. For the case where the phase step is $\pi/2$ between speakers, the power spectrum has a shape

$$dP(\Delta f) \propto \frac{\sin^2(M\pi d / 2u)(\Delta f + u / 2d)}{\sin^2(\pi d / 2u)(\Delta f + u / 2d)}, \tag{4.18}$$

which peaks at $\Delta f = -u / 2d$ and has a spectral width of $2u / Md$ from the peak to the first null. The spectral power is shown in Figure 4.32 for the case $u = 5$ m s^{-1}, $kd = 7$, $M = N = 8$, and $d(\partial\varphi / \partial x) = \pi / 2$. The comparable bistatic case, with the transmitter–receiver separation of 50 m, a vertical transmitted beam, and a tilted receiver beam aimed at a height of 50 m, is shown in Figure 4.33. The effect of unshaded lobes in this case is much more damaging. Note also the bias in the wind speed estimated from the spectral peak position. For these reasons it is very desirable to use a pulsed system with a bistatic configuration.

4.6   CONTINUOUS AND PULSED SYSTEMS

As discussed above, a monostatic system *must* use a transmitted pulse, since the same transducer cannot transmit and receive simultaneously. Even if separate transducers were used for transmitting and receiving within a single antenna, the high-power transmitted sound would overload the sensitive receiver. We have also seen that for a bistatic system it is *desirable* to use a transmitted pulse so that the scattering volume is better defined spatially.

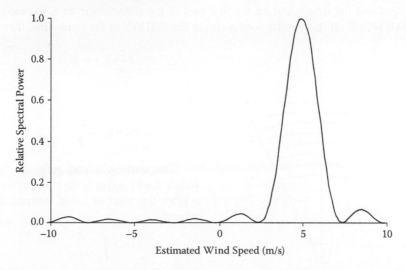

FIGURE 4.32 The relative power spectrum, shown with a wind-speed axis, for a finite monostatic beam.

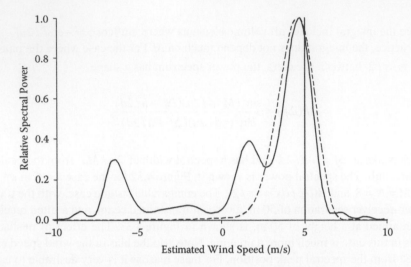

FIGURE 4.33 As for Figure 4.32, but for the bistatic case, with transmitter–receiver separation of 50 m, a vertical transmitted beam and a tilted receiver beam aimed at a height of 50 m. Unshaded case (solid line); shaded case (dashed line).

For a system having pulse duration τ and with speed of sound c, the pulse is spread over a height range of $c\tau$, as shown in Figure 4.34.

However, sound is detected at any one time from a volume having height range $c\tau/2$. This can be seen by considering reflections from two vertically separated turbulent patches, as shown in Figure 4.35.

The time taken for the last part of the reflection from z_1 to reach the SODAR is $2z_1/c+\tau$ and the time taken for the first part of the reflection from z_2 to reach the SODAR is $2z_2/c$. If these reflections arrive at the SODAR at the same time, then

$$2\frac{z_1}{c}+\tau=2\frac{z_2}{c}$$

or

$$z_2-z_1=\frac{c\tau}{2}.$$

The earliest sound reflected from height z will arrive at the receiver a time t after the start of pulse transmission, where

$$t=2\int_0^z \frac{\mathrm{d}z}{c(z)}.$$

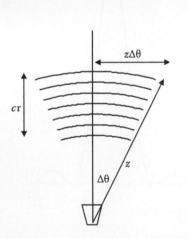

FIGURE 4.34 The geometry of a transmitted pulse in a conical beam of half-angle $\Delta\theta$.

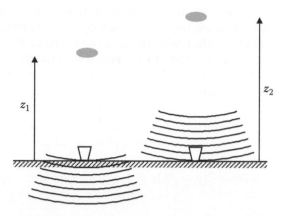

FIGURE 4.35 Sound received from two vertically separated scatterers.

It is usually assumed that c is constant with height, and so the height *estimated* from the sound speed is

$$\hat{z} = \frac{c(z=0)t}{2}.$$

(4.19)

How good is this approximation? As shown in Chapter 3, sound speed varies with temperature and humidity, from which we could write

$$
\begin{aligned}
t &= 2\int_0^z \left[\frac{1}{c(0)} + \left(\frac{d}{dz}\frac{1}{c}\right)_{z=0} z + \cdots \right] dz \\
&= 2\int_0^z \left[\frac{1}{c(0)} + \left(-\frac{1}{c^2}\frac{dT}{dz}\frac{dc}{dT}\right)_{z=0} z + \cdots \right] dz \\
&= 2\int_0^z \left[\frac{1}{c(0)} + \left(-\frac{1}{c^2}\frac{dT}{dz}\frac{c}{2T}\right)_{z=0} z + \cdots \right] dz \\
&= \frac{2z}{c(0)}\left[1 - \frac{1}{4T}\frac{dT}{dz}z + \cdots \right].
\end{aligned}
$$

Since $T \approx 300$ K and a typical lapse rate is 1°C per 100 m, the first correction term is of order $10^{-5}z$ for z in meters, or a correction of only 1% in 1 km, which can generally be ignored.

The acoustic power is spread over a larger area, $\pi(z\Delta\theta)^2$, as height increases. Sound is therefore scattered back from a volume

$$V = \frac{\pi(z\Delta\theta)^2 c\tau}{2}.$$

(4.20)

The strength of scattering depends on the turbulence levels and can be represented by a differential scattering cross-section σ_s. This scattering cross-section is the equivalent area of a perfect reflector, per unit volume of air, and per unit solid angle of scattering (the solid angle of scattering is the effective sensing area of the receiver, divided by the square of the distance z). This means that the intensity of scattered sound is proportional to $\sigma_s V$ or to $\sigma_s c\tau / 2$. Generally σ_s is a function of height z, so $\sigma_s c\tau / 2$ should be replaced with $\displaystyle\int_{z_1}^{z_2} \sigma_s \, dz$. Also, if the pulse amplitude is modulated by $m(t)$, then the intensity of scattered sound depends on

$$\int_{z_1}^{z_2} \sigma_s(z) m\left(\frac{2}{c}[z_2 - z]\right) dz = \int_{z_2-c\tau/2}^{z_2} \sigma_s(z) m\left(\frac{2}{c}[z_2 - z]\right) dz.$$

The echo power at the receiver is therefore proportional to the *convolution* of turbulent cross-section profile and pulse amplitude.

$$P_R(t) \propto \int_{c(t-\tau)/2}^{ct/2} \sigma_s(z) m\left(t - \frac{2z}{c}\right) dz.$$

Note that this equation predicts a continuous time record, with information at time t coming from throughout the height range $\frac{1}{2}c(t-\tau) \le z \le \frac{1}{2}ct$. This time record is interpreted as a height profile to give

$$P_R(\hat{z}) \propto \int_{\hat{z}-c\tau/2}^{\hat{z}} \sigma_s(z) m\left(\frac{2[\hat{z}-z]}{c}\right) dz.$$

$$(4.21)$$

What is the spatial resolution of a pulsed system? The Fourier transform of (4.21) with respect to a spatial wavenumber κ gives

$$P(\kappa) = \int_{-\infty}^{\infty} P_R(\hat{z}) e^{-j\kappa\hat{z}} \, d\hat{z}$$

$$\propto \int_{-\infty}^{\infty} \int_{\hat{z}-c\tau/2}^{\hat{z}} \sigma_s(z) m\left(\frac{2[\hat{z}-z]}{c}\right) dz\, e^{-j\kappa\hat{z}} \, d\hat{z}$$

$$\propto \int_{-\infty}^{\infty} \sigma_s(z) \left[\int_{-\infty}^{\infty} m\left(\frac{2[\hat{z}-z]}{c}\right) e^{-j\kappa\hat{z}} \, d\hat{z}\right] dz$$

$$\propto \int_{-\infty}^{\infty} \sigma_s(z) \left[\frac{c}{2} e^{-j(c\kappa/2)t} \int_{-\infty}^{\infty} m(\hat{t}-t) e^{-j(c\kappa/2)(\hat{t}-t)} \, d\hat{t}\right] dz$$

$$\propto \int_{-\infty}^{\infty} \sigma_s(z) \left[\frac{c}{2} e^{-i(c\kappa/2)t} M\left(\frac{c\kappa}{2}\right)\right] dz$$

$$\propto \frac{c}{2} M\left(\frac{c\kappa}{2}\right) \int_{-\infty}^{\infty} \sigma_s(z) e^{-j\kappa z} \, dz$$

or

$$P(\kappa) \propto \frac{c}{2} M\left(\frac{c\kappa}{2}\right) S(\kappa),$$

$$(4.22)$$

where the change of limits on the $\hat{z}$ integral is possible providing that the finite nature of the pulse is included in the pulse Fourier transform M. This equation shows that the spatial structure of the atmosphere, represented by $S(\kappa)$, is filtered by the pulse characteristics represented in spatial terms by $M(c\kappa/2)$. For example, a simple square pulse

$$m(t) = \begin{cases} 0 & \text{for } t < 0, \\ 1 & \text{for } 0 \leq t \leq \tau, \\ 0 & \text{for } \tau < t \end{cases}$$

gives

$$\left| M\left(\frac{c\kappa}{2}\right) \right| = \frac{\sin(c\kappa\tau/4)}{(c\kappa\tau/4)}.$$

This clearly has oscillations, which means that spatial range of sensitivity is broad. A better pulse shape, or modulation, is the "Hanning window"

$$m(t) = \begin{cases} 0 & \text{for } t < 0, \\ \frac{1}{2}\left(1 - \cos 2\pi \frac{t}{\tau}\right) & \text{for } 0 \leq t \leq \tau, \\ 0 & \text{for } \tau < t, \end{cases}$$

which gives

$$\left| M\left(\frac{c\kappa}{2}\right) \right| = \frac{1}{1 - (c\kappa\tau/4\pi)^2} \frac{\sin(c\kappa\tau/4)}{(c\kappa/4)}.$$

These two filter functions are shown in Figure 4.36. It can be seen that the Hanning pulse has reduced ripples at the expense of being wider.

The effect of this spatial filtering can also be illustrated through a simple model for $\sigma_s(z)$. Assume the spatial structure $S(\kappa)$ is band-limited to $\kappa < \kappa_{max}$ and has a Gaussian probability at each κ. Figure 4.37 shows two randomly generated profiles $\sigma_s(z)$ and the corresponding $P_R(\hat{z})$ profiles using a Hanning window and a pulse with $c\tau/2 = 10$ m. In the left-hand plot, $S(\kappa)$ is limited to $0 \leq \kappa \leq 0.5(4\pi/c\tau)$, and the spatial resolution of the pulse is close to the smallest spatial scales of the model atmosphere. In this case, the retrieved profile is close to the original. In the right-hand plot $S(\kappa)$ is limited to $0 \leq \kappa \leq 3(4\pi/c\tau)$, and the spatial resolution of the pulse is much larger than the smallest spatial scales of the model atmosphere. In this case, the retrieved profile shows large variations from the original. In both cases, residuals using a square pulse are about twice as large.

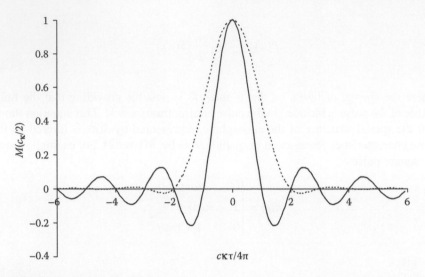

FIGURE 4.36 Pulse spatial filter functions for a square pulse (solid line) and a Hanning pulse (dashed line).

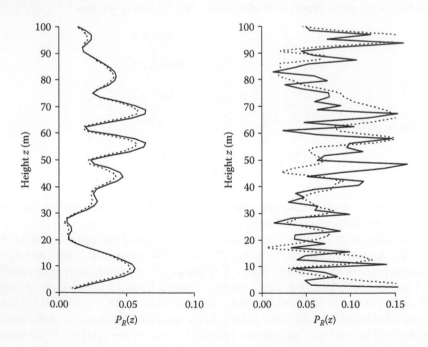

FIGURE 4.37 Profiles of $\sigma_s(z)$ (solid line) and $P_R(\hat{z})$ (dashed line) for a 10-m Hanning pulse. In the left-hand plot $0 \leq \kappa \leq 4\,\pi\,\mathrm{p}/2c\tau$ and in the right-hand plot $0 \leq \kappa \leq 12\,\pi\,\mathrm{p}/c\tau$.

The above simply emphasizes that the pulse should ideally be of a spatial length to match the spatial scales of interest in the atmosphere. There is no solid rule for what this length should be, but this question will be discussed in more detail later.

Note that M is just the frequency spectrum of the pulse modulation m, and can be thought of as a spatial spectrum or a frequency spectrum (since time and height are analogues here). When Doppler frequency shifts need to be estimated to obtain wind components, it is again undesirable to have the modulation spectrum oscillating, and the Hanning shape is advantageous. Again the width of M in frequency space should be smaller than the frequency structure of interest. For estimating the Doppler shift, the width of M should be as small as possible. However, this implies a long pulse. The compromises required will be discussed later.

4.7   GEOMETRY OF SCATTERING

Figure 4.21 and Eq. (4.11) give the geometry for the general bistatic case. The scattering angle is

$$\beta = \cos^{-1}\left[\cos\theta_R\left(\frac{D}{z}\sin\theta_T\cos\phi_R - \frac{1}{\cos\theta_T}\right)\right],$$

where D is the separation between the transmitter and the receiver, z the height of scattering, θ_R and θ_T are the zenith angles of the receiver and transmitter beams, and ϕ_R the azimuth angle of the receiver beam measured with respect to the line joining the transmitter and the receiver. For a monostatic SODAR, $D = 0$ and $\theta_R = \theta_T$, so $\beta = \pi$. For a bistatic SODAR having vertical transmitter beam, $\theta_T = 0$, so $\beta = \pi - \theta_R$. As seen in Chapter 3, the scattered intensity is proportional to

$$\Phi_n(\kappa) = \cos^2\beta\left[\cos^2\left(\frac{\beta}{2}\right)\frac{\Phi_V(\kappa)}{c^2} + \frac{\Phi_T(\kappa)}{4T^2}\right],$$

where k is the wavenumber of the sound and the Bragg wavenumber is

$$\kappa = 2k\sin(\beta/2).$$

The function $\Phi_T(\kappa)$ is closely related to the autocorrelation function for temperature fluctuations, which is a component of C_T^2 as seen in Chapter 2. Some angular dependence should be present, so by analogy with Rayleigh scattering, which is also weak scattering, one might expect σ_s to depend on k^4. Intuitively, then, one might write this part of the scattering cross-section in the form

$$\sigma_s = A\frac{C_T^2}{T^2}k^4\kappa^q,$$

where q and A are dimensionless constants. Dimensional analysis shows that $q = -11/3$, so

$$\sigma_s = A\frac{k^4}{\kappa^{11/3}}\frac{C_T^2}{T^2}. \tag{4.23}$$

The total frequency dependence from scattering is very weak ($f^{1/3}$). When the 3D acoustic wave equation is solved exactly for turbulent scattering

$$\sigma_s = \frac{1}{8}\frac{k^4}{\kappa^{11/3}}\cos^2\beta\left[0.033\frac{C_T^2}{T^2}+0.76\cos^2\left(\frac{\beta}{2}\right)\frac{C_V^2}{\pi c^2}\right], \tag{4.24}$$

where

$$\kappa = 2k\sin\frac{\beta}{2}. \tag{4.25}$$

4.8   THE ACOUSTIC RADAR EQUATION

For a monostatic SODAR, the acoustic power is spread over a larger area, $\pi(z\theta)^2$, as height increases, so the *sound intensity* (power per unit area) decreases with height. Similarly, sound reflected downward from turbulence also spreads out, so that received intensity is proportional to z^{-4}. This is one of the main reasons for range limitations in SODARs.

An electrical power P_T is dissipated during pulse transmission. The antenna converts this into acoustical power $P_T G$ in the upward beam, with some power lost into other directions and in heat. At height z, sound of intensity $P_T G / \pi(z\theta)^2$ falls on turbulent patches in the scattering volume. The patches reflect some power back toward the antenna, equivalent to them being perfect reflectors of area σ_s per unit volume of air and per unit solid angle of scattering. The reflected power is therefore $P_T G\sigma_s V / \pi(z\theta)^2$ per unit solid angle. If the receiving antenna has an area A_e and is distance z from the turbulent volume, the solid angle is A_e/z^2. The power received is therefore $P_T GA_e\sigma_s V / \pi\theta^2 z^4$. The area A_e is really an *effective area* since it includes the conversion by the antenna microphones of acoustic energy into electrical energy.

The absorption over the travel distance of $2z$ also needs to be allowed for. The power absorbed by the air in a small distance depends on the incident power, an absorption coefficient (which allows for the absorbing properties of air), and on the distance. The result is an exponential fall off with distance, so the received electrical power is

$$P_R = P_T GA_e\sigma_s\frac{c\tau}{2}\frac{e^{-2\alpha z}}{z^2}. \tag{4.26}$$

The terms in (4.26) can be grouped into instrumentation terms:

1. Transmitted power P_T,
2. Antenna transmitting efficiency G,
3. Antenna effective receiving area A_e,

4. Pulse duration τ,
   and atmospheric terms:
5. Height z,
6. Absorption of air α,
7. Turbulent scattering cross-section σ_s.

Calibration is aimed at measuring the combined instrumentation terms.

More generally, if the pulse amplitude is modulated by $m(t)$, then

$$P_R(t) \approx P_T GA_e \frac{e^{-\alpha ct}}{(ct/2)^2} \int_{c(t-\tau)/2}^{ct/2} \sigma_s(z)m\left(t - \frac{2z}{c}\right)dz, \tag{4.27}$$

where it has been assumed that absorption is not height-dependent and the spherical spreading has been taken outside the height integral.

For a pulsed bistatic configuration with the receiver distance D from the vertical-beam transmitter, (4.26) is modified to

$$P_R = P_T GA_e(\pi - \beta)\sigma_s(\beta)\frac{c\tau}{2}\frac{e^{-\alpha(z+R_R)}}{R_R^2}, \tag{4.28}$$

where the angular dependence of σ_s and A_e must be allowed for.

4.9   ACOUSTIC BAFFLES

Although commercial SODARs have been marketed as an array of speakers without surrounding acoustic shielding, these systems have not been successful.

The two major non-random noise sources for a SODAR are

* reflections of the pulse from non-atmospheric objects ("fixed echoes" or "clutter") and
* acoustic noise from vehicles, animals, trees, etc., reaching the antenna.

Both these noise sources can be greatly reduced by surrounding the SODAR antenna with an acoustic baffle.

It is worth considering how small the sound intensity needs to be, if reflected off a perfectly reflecting object, so as not to compete with the turbulence signal. For backscatter

$$\sigma_s = 6\times10^{-4}\lambda^{-1/3}\frac{C_T^2}{T^2}.$$

Typically $C_T^2 = 10^{-4}$ to 10^{-3} K^2 $m^{-2/3}$ and $\lambda = 0.1$ m, so $\sigma_s \sim 10^{-11}$ m^{-1} sr^{-1}. As a measure of efficiency, for a beam half-angle of $\pi/30$ and at $z = 100$ m, the cross-section area of a typical SODAR scattering volume would be $\pi(100\pi/30)^2 = 340$ m^2, and the scattering volume would be $8.5 \times 340 = 2890$ m^3. So the *physical* scattering area per unit volume is of order 0.1 m^{-1}. This emphasizes how weak is the scattering by turbulence. The received power from turbulent scattering is very small. For

$P_T = 10$ W, $GA_e = 2$ m^2, $c\tau/2 = 8.5$ m, $\sigma_s = 10^{-11}$ m^{-1} sr^{-1}, $z = 100$ m, and absorption loss 2 dB, the received power is of order 10^{-13} W. If a hard object at low elevations is distance r away, the intensity at the object will be $G_L P_T/[\pi(x\theta_L)^2]$, where the subscript L has been used to indicate a side lobe angle. After reflection and traveling back the same distance to the antenna, the intensity will decrease by a factor of 2. If the area of the antenna is A, and side lobe energy is gathered with the same efficiency as it is transmitted, then the received fixed echo power is

$$P_L = P_T G_L^2 A \frac{e^{-2\alpha x}}{2\pi(x\theta_L)^2}.$$

The ratio of side lobe received power to turbulence received power from the same distance is

$$\frac{P_L}{P_R} = \left(\frac{G_L}{G}\right)^2 \frac{1}{c\tau\sigma_s\pi(\theta_L)^2}.$$

For $c\tau = 20$ m, $\sigma_s = 10^{-11}$ m^{-1} sr^{-1}, and $\theta_L = 20°$, P_L and P_R are comparable if G_L/G = −50 dB. Roughly speaking, this means that the intensity of side lobes must be 50 dB smaller than the intensity of the main beam. Unshielded dish antennas and shaded phased array antennas come close to this benchmark, so an acoustic baffle does not need to have extremely good transmission loss. The real problem occurs because of diffraction over the top edge of the baffle.

The ideal baffle material should make an absorbing, non-reflecting and non-transmitting shield. Generally, fiberglass or marine plywood coated on the inside with acoustic foam is used. Some baffles have thin lead sheet glued to the fiberglass or wood substrate. This construction is usually sufficient to stop any significant sound energy penetrating *through* the baffle walls.

The baffle will be generally conical with the main SODAR lobe emerging from its top. Most baffles have straight sides, although some earlier SODARs used horn-shaped baffles in which the top curved away from the SODAR vertical axis. This design is presumably based on the use of horns to couple small speaker drivers to the atmosphere. However, the application is much different from speaker horns, since the scales are much larger than a wavelength and acoustic impedance matching is not relevant. There is likely to be some benefit, though, in having the rim of the baffle absorbing (this effectively reduces the aperture weighting at the rim) and an outwardly curved rim will aid this.

For phased arrays, the baffle needs to have a wider exit so that tilted beams do not intersect the baffle edges too much. The top rim of the baffle will be in the near field of the SODAR beam (rays from different parts of the antenna to a point on the rim will not be parallel). However, detailed calculations have shown that far-field approximations are generally sufficient to optimize a design. Baffles can have a circular cross-section or be some polygon: for the following we just consider the rim as if it were a circle.

The baffle acts as another circular hole with its own circular diffraction pattern. The aperture weighting function in this case is the beam pattern at the rim level. Of prime importance is the acoustic intensity near the rim itself. If the aperture weighting function were uniform across the baffle rim level, then the angular intensity pattern would be

$$\left[\frac{2J_1(ka\sin\theta^*)}{ka\sin\theta^*}\right]^2$$

$$(4.29)$$

with a being the radius of the rim of the baffle and θ^* defined as in Figure 4.38.

The result is that as the baffle is made higher, the zenith angle to the baffle rim becomes smaller and the acoustic intensity at the rim is generally larger. On the other hand, the diffracted angle θ^* toward a given direction is also larger. Larger diffraction angle means a lower diffracted intensity. So the height of a baffle is a compromise between the two competing diffraction effects from the antenna and the baffle (and of course the desire to have a less bulky instrument!).

For example, assume that the diffracted sound which causes the fixed echoes travels horizontally, so $\theta^*= 90°-\theta$. The combination of beam intensity at the rim and the diffracted intensity is

$$\left[\frac{2J_1(ka\sin\theta)}{ka\sin\theta}\right]^2\left[\frac{2J_1(ka\sin\theta^*)}{ka\sin\theta^*}\right]^2.$$

This is plotted in Figure 4.39 versus h, the height of the baffle, assuming $a = 0.6$ m and $f_T = 3400$ Hz. Also shown is the case where Hanning shading has been used across the antenna. First, it is clear that the effect of increased baffle height (and more intercepted beam) is almost cancelled by the increased diffraction angle: there is no advantage to having a high baffle. This explains why some manufacturers (e.g., Scintec) prefer low baffles. Secondly, the effect of antenna shading is dramatic!

This example, with only very approximate numbers, emphasizes the need for caution in designing SODAR antennas and baffles. Diffraction effects can also be minimized by choosing the baffle height so that the minimum in the array pattern falls on the rim of the baffle. This approach is similar to the reduction in side lobes for a dish antenna caused by having a directional speaker mounted at its focus. The formula above for phased array beam patterns suggests that the intensity goes to

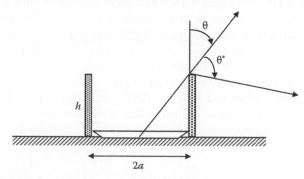

FIGURE 4.38 Baffle geometry.

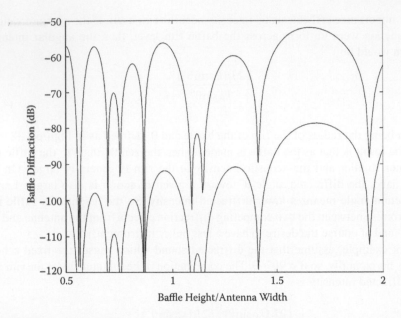

FIGURE 4.39 The intensity of sound diffracted from the rim of a baffle, without shading (upper curve), and with Hanning antenna shading (lower curve).

zero at the minima. In practice, near-field effects, finite speaker size, inequalities in speaker characteristics, and other effects mean that measured minima are not zero. This design requires baffle height h and rim radius a to be related to the antenna pattern minimum angle via $a/h = \tan \theta_{min}$.

In practice this is also not entirely possible since the beam is tilted at times and the user might also change the transmitted frequency. So most SODAR baffle manufacture is still based on trial and error.

Finally, the rim of the baffle can be broken up by giving it a ragged shape, usually having triangular peaks. In the first investigation of this, by Wirt (1979), the name "thnadner" was used, borrowed from a popular children's story by "Dr Zeus." A more in-depth study by Werkhoven and Bradley (1997) showed that this rim modulation could give more than 6 dB reduction in diffraction (Fig. 4.40) and also entrenched the name thnadner. Several manufacturers use baffles with thnadners, including the relatively recently available AQ500 (see Fig. 4.41). The actual height and width of the triangular peaks do not seem critically related to wavelength, so that rather than a resonant phenomenon, it is more likely to be creating a "soft edge" effect by gradually changing the percentage of rim.

In addition, Werkhoven and Bradley (1997) confirmed that the least edge diffraction occurs when the rim of the baffle lies at the first zero of the $J_1(ka \sin \theta)$ function. This is when $ka \sin \theta = 3.83$. This essentially defines the aspect of the cylindrical part of the SODAR baffle. With some rearrangement

$$\left(\frac{h}{a}\right)^2 = \left(\frac{ka}{3.83}\right)^2 - 1. \tag{4.30}$$

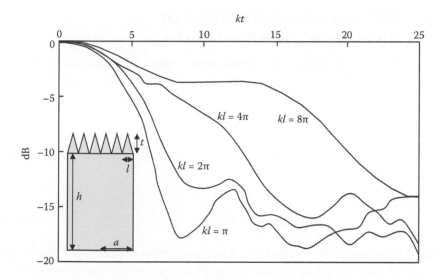

FIGURE 4.40 The extra suppression of diffraction side lobes due to thnadners of height t and spacing l on a cylindrical baffle of height $h = 86/k$ and radius $a = 24/k$.

FIGURE 4.41 The AQ500 SODAR with thnadners.

4.10   FREQUENCY-DEPENDENT FORM OF
THE ACOUSTIC RADAR EQUATION

The acoustic radar equation (4.24) does not show Doppler shift information. Assume the emitted signal, $p_T(t)$, is a cosine pulse of constant frequency f_T and duration τ, weighted by an envelope $m(t)$:

$$p_T(t) = m(t)\exp(j2\pi f_T t). \tag{4.31}$$

The received signal, $p_R(t)$, will then have a height-dependent phase shift and a Doppler-modified frequency

$$f_D = \left(1 - 2\frac{V}{c}\right)f_0 = \xi f_T, \tag{4.32}$$

where ξ is the ratio of the received signal frequency to transmitted signal frequency. For typical atmospheric absorption values of around 10^{-2} m^{-1} the attenuation is approximately constant over one range gate, especially for increasing heights. The received signal is then given by

$$p_R(t) \propto \int_0^\infty \sigma_s(z)m[\xi(t-t_z)]\exp[j\xi 2\pi f_T(t-t_z)]\mathrm{d}z. \tag{4.33}$$

For existing instruments the pulse is generally very short ($\tau \approx 50$ ms) and the reflected signal is divided into $n = 1, 2, \dots, N$ sequences of approximate duration τ, each corresponding to a certain height range, or range gate extent, $c\tau/2$, of about 10 m. The wind speed is approximated by a constant, i.e. $\xi(t, z) \approx \xi$, representing the mean wind speed of the 10-m-thick measuring volume.

For a Gaussian envelope $m(t) = \exp[-(1/2)(t/\sigma_m)^2]$ the Fourier transform of the reflected signal can be written as

$$P_R(f) \propto \int_{-\infty}^\infty \mathrm{d}t \int_{c(t-\varsigma)/2}^{ct/2} \sigma_s(z)m[\xi(t-t_z)]\exp\{j2\pi[f_T\xi(t-t_z) - ft]\}\,\mathrm{d}z$$

$$= \int_{c(t-\varsigma)/2}^{ct/2} \sigma_s(z)\exp(-2j\xi kz)\,\mathrm{d}z \int_{-\infty}^\infty \exp\left[-j2\pi(f - \xi f_T)t - \frac{\xi^2(t-t_z)^2}{2\sigma_m^2}\right]\mathrm{d}t$$

$$= \frac{1}{2\sqrt{2\pi}\sigma_f}\exp\left[-\frac{(f - f_D)^2}{2\sigma_f^2}\right]S(\kappa_D), \tag{4.34}$$

where the definition $\sigma_f = \xi/2\pi\sigma_m$ has been used. This spectrum is a product of a Gaussian and the spatial Fourier transform, $S(\kappa_D)$ of the scattering cross-section, σ_s, at scales $\kappa_D = 2\xi k$ where $k = 2\pi f_T/c$ is the original wavenumber. From Eq. (4.34) the power spectrum $|P_R(f)|^2$ can be derived as

$$|P_R(f)|^2 = \frac{1}{8\pi\sigma_f^2} |S(\kappa_D)|^2 \exp\left|-\frac{(f-f_D)^2}{\sigma_f^2}\right|, \qquad (4.35)$$

which peaks at $f = \xi f_T = f_D$. From the peak position, f_D, and Eq. (4.15), the mean wind speed within each range gate can be calculated. Strictly speaking, ξ is a function of time and height, $\xi = \xi(t,z)$, as it represents a turbulent wind speed, and the peak around f_D will be broadened.

Examination of Eq. (4.34) also shows that, if the envelope is constant, the received signal can be written in the form

$$p_R(t) \propto S(\kappa_D)\exp(-j2\pi f_D t). \qquad (4.36)$$

Generally, the refractive index fluctuations are very small and the only significant reflectivity occurs when the back-scattered signals from many turbulent patches interfere constructively. This "Bragg" condition essentially Fourier analyses the scattering volume at spatial periodicities of half-wavelength size within the continuous $S(\kappa)$ spectrum. This is evidenced in the appearance of a preferred scale size κ_D.

For a Gaussian envelope, $m(t)$, with $\tau = 50$ ms and standard deviation $\sigma_m = \tau/6$, the power spectral peak has a Gaussian shape with a standard deviation of around $\sigma_f / \sqrt{2} = 13.5$ Hz. From Eq. (4.29), and for a carrier frequency, f_T, between 3 and 4 kHz, $V = 0.5$ to 1 m s^{-1} would give a shift $f_D - f_T$ comparable to 13.5 Hz, so the spectral resolution is typically poor in comparison with expected Doppler shifts. For this reason, and also in order to discriminate the signal against environmental noise, a SODAR measurement generally consists of an average of some tens of individual measurements to obtain a more reliable estimate of f_D. This means that the wind speed is typically measured over a period of around five or ten minutes and thus represents a temporal average over this time.

4.11   OBTAINING WIND VECTORS

In the monostatic case, Eq. (4.16) gives the Doppler shift, Δf, for a single beam as a linear combination of the components (u, v, w) of wind vector V, which can be rewritten in the form

$$-\frac{c}{2}\frac{\Delta f}{f_T} = (\sin\theta\cos\phi)u + (\sin\theta\sin\phi)v + (\cos\theta)w. \qquad (4.37)$$

The comparable equation, for the bistatic configuration shown in Figure 4.29, is

$$c\frac{\Delta f}{f_T} = (\sin\theta\cos\phi)u + (\sin\theta\sin\phi)v + (-1-\cos\theta)w. \qquad (4.38)$$

At least three such equations are required to find u, v, and w at each range gate height. Typically, for monostatic systems, three or five beams are used, as depicted in Figure 4.42, whereas Figure 4.43 shows a comparable 3-beam bistatic system, with beam 3 being monostatic.

There are three or five spectral peak positions linearly related to the three velocity components as follows:

$$
\begin{pmatrix} \eta_1 \\ \eta_2 \\ \eta_3 \\ \eta_4 \\ \eta_5 \end{pmatrix} = \begin{pmatrix} -\dfrac{c}{2}\dfrac{\Delta f_1}{f_T} \\ -\dfrac{c}{2}\dfrac{\Delta f_2}{f_T} \\ -\dfrac{c}{2}\dfrac{\Delta f_3}{f_T} \\ -\dfrac{c}{2}\dfrac{\Delta f_4}{f_T} \\ -\dfrac{c}{2}\dfrac{\Delta f_5}{f_T} \end{pmatrix} = \begin{pmatrix} \sin\theta\cos\phi & \sin\theta\sin\phi & \cos\theta \\ -\sin\theta\sin\phi & \sin\theta\cos\phi & \cos\theta \\ 0 & 0 & 1 \\ -\sin\theta\cos\phi & -\sin\theta\sin\phi & \cos\theta \\ \sin\theta\sin\phi & -\sin\theta\cos\phi & \cos\theta \end{pmatrix} \begin{pmatrix} u \\ v \\ w \end{pmatrix}
$$

$$
= \begin{pmatrix} \sin\theta & 0 & \cos\theta \\ 0 & \sin\theta & \cos\theta \\ 0 & 0 & 1 \\ -\sin\theta & 0 & \cos\theta \\ 0 & -\sin\theta & \cos\theta \end{pmatrix} \begin{pmatrix} \cos\phi & \sin\phi & 0 \\ -\sin\phi & \cos\phi & 0 \\ 0 & 0 & 1 \end{pmatrix} \begin{pmatrix} u \\ v \\ w \end{pmatrix}
$$

(4.39)

or

$$ N = BRV \tag{4.40} $$

where V is a 3×1 vector, R is a 3×3 matrix giving rotation by angle ϕ around the z-axis, and B is a 5×3 (or 3×3 for a 3-beam system) beam-steering matrix. The comparable beam-steering matrix for a 5-beam bistatic SODAR using the notation in (4.38) is

$$
B = \begin{pmatrix} \sin\theta & 0 & -1-\cos\theta \\ 0 & \sin\theta & -1-\cos\theta \\ 0 & 0 & -2 \\ -\sin\theta & 0 & -1-\cos\theta \\ 0 & -\sin\theta & -1-\cos\theta \end{pmatrix}.
$$

(4.41)

For 3-beam systems, the system is solved by

$$ \hat{V} = (BR)^{-1}N \tag{4.42} $$

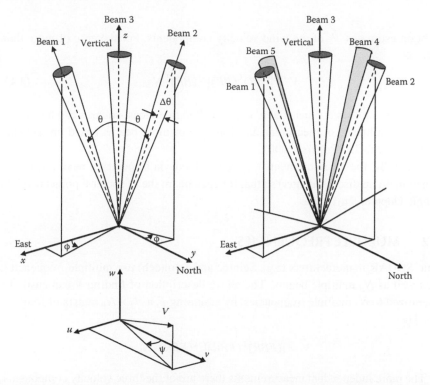

FIGURE 4.42 Typical beam geometries for a 3-beam and a 5-beam monostatic system, showing the relationship to wind components. The transmission in a particular beam direction, and the analysis of the echoes from that beam, is usually completed before transmission in the next beam direction.

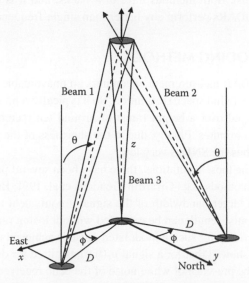

FIGURE 4.43 A possible 3-beam bistatic SODAR geometry, where the vertical beam (beam 3) is monostatic.

to obtain estimates $\hat{V}$ of the wind velocity components. For the case of more than 3 beams

$$\hat{V} = [(BR)^{\mathrm{T}}(BR)]^{-1}(BR)^{\mathrm{T}} N \qquad (4.43)$$

obtains least-squares estimates of u, v, and w. Weighted least-squares can be used to reduce the influence of measurements from one or more beams based on software consistency checks (Rogers, 2000).

Generally, the "measurements" N are obtained via a spectrum peak estimation algorithm (to be discussed later) so that for each beam the M spectral points result in a single Doppler shift value.

4.12   MULTIPLE FREQUENCIES

Some SODAR manufacturers (e.g., Scintec and Remtech) use multiple frequencies, f_T, as well as N_B multiple beams. The above description of finding V can easily be generalized to N_T multiple frequencies, by assigning F, a $N_T \times N_B$ matrix of frequencies. Then

$$\hat{V} = [(FBR)^{\mathrm{T}}(FBR)]^{-1}(FBR)^{\mathrm{T}} N.$$

The more independent measurements there are of the three velocity components, the greater accuracy would be expected. However, multiple frequencies must either be transmitted together (generally resulting in reduced power output at each individual frequency) or be transmitted sequentially (resulting in slower cycle time around the individual beams). Both methods have drawbacks, and it is still not clear that multi-frequency SODARs perform any better than single-frequency SODARs.

4.13   PULSE CODING METHODS

A conventional SODAR measurement encounters an unavoidable trade-off between height and frequency (wind speed) resolution. So it typically represents a wind speed measurement averaged over a height range of around ten meters and over a time interval of up to ten minutes. Finally, due to the shortness of the pulse the received signal energy and thus the SNR is very low.

To help overcome these limitations, there have been several proposals for use of pulse compression methods (e.g., Girardin-Gondeau et al. 1991; Bradley 1999). Such methods result in a larger bandwidth of the signal – equivalent to that of a shorter pulse – so that the pulse length can be extended without losing range resolution.

Echo signal recovery for such a modulated pulse is achieved through a matched filter. North (1963) showed that for a signal $p_T(t)$ transmitted for duration τ, the optimum detection in the presence of white noise of the echo received time t_z later from a non-moving point target at distance z is obtained using a receiver having a matched filter output

$$\int_0^\tau p_T(t)p_T(t+t_z-\Delta t)dt, \tag{4.44}$$

which is a time-delayed autocorrelation function of $p_T(t)$. However, when the radial velocity, V, of the target is significant compared with the signal propagation speed, c, the Doppler shifted reflections are no longer replicas of the transmitted signal. In this case an ambiguity function, equal to the squared magnitude of the Doppler-shifted correlation

$$\Psi(\Delta t, f_D) = \left| \int p_T(t)p_T(t-\Delta t)\exp(j2\pi f_D t)dt \right|^2 \tag{4.45}$$

is used to describe how closely the matched filter response estimates the target position (through delay Δt) and radial velocity (through Doppler frequency f_D). In Eq. (4.45), Δt is measured relative to the target echo signal time delay and f_D is measured relative to the Doppler frequency from the target, so non-zero values for either parameter represent an error in parameter estimation using the matched filter.

To overcome mismatch between filter and echo signal, the filter can include a variable Doppler shift. Adjusting this variable Doppler shift to give either maximum amplitude of the autocorrelation function in (4.45) or to give minimum phase drift between echo signal and filter will give an estimate of the Doppler shift.

Both frequency-encoding and phase-encoding are common approaches. For example, a step-chirp signal has a transmitted signal of the form

$$p_T(t) = \exp[j2\pi(f_0+m\Delta f)t], \quad \frac{m}{M}\tau \le t < \frac{m+1}{M}\tau, \quad m=0,1,\dots,M-1$$

as shown in Figure 4.44.

The received signal, within a range gate at z, is proportional to

$$p_R(t) = \sigma_s(z)\exp[j2\pi\xi(f_0+m\Delta f)(t-t_z)], \quad \frac{m}{M}\frac{\tau}{\xi}\le t-t_z < \frac{m+1}{M}\frac{\tau}{\xi}, \quad m=0,1,\dots,M-1.$$

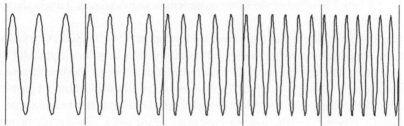

FIGURE 4.44 A step-chirp signal having $M=5$ frequency steps.

The matched filter output is proportional to

$$\Psi(t_z,\xi) = \sigma_s(z) \sum_{m=0}^{M-1} \int_{\frac{m}{M}\frac{\tau}{\xi}}^{\frac{m+1}{M}\frac{\tau}{\xi}} \exp\left[j2\pi\xi\left(f_0+m\Delta f\right)t'\right]\exp\left[-j2\pi\hat{\xi}\left(f_0+m\Delta f\right)t'\right]dt'$$

$$= \sigma_s(z) \sum_{m=0}^{M-1} \int_{\frac{m}{M}\frac{\tau}{\xi}}^{\frac{m+1}{M}\frac{\tau}{\xi}} \exp\left[j2\pi\left(f_0+m\Delta f\right)\left(\xi-\hat{\xi}\right)t'\right]dt'$$

$$= \sigma_s(z) \sum_{m=0}^{M-1} \frac{\exp\left[j2\pi\left(f_0+m\Delta f\right)\left(\xi-\hat{\xi}\right)\frac{m+1}{M}\frac{\tau}{\xi}\right] - \exp\left[j2\pi\left(f_0+m\Delta f\right)\left(\xi-\hat{\xi}\right)\frac{m}{M}\frac{\tau}{\xi}\right]}{j2\pi\left(f_0+m\Delta f\right)\left(\xi-\hat{\xi}\right)}$$

$$= \frac{\sigma_s(z)}{M}\frac{\tau}{\xi} \sum_{m=0}^{M-1} \exp\left[j2\pi\left(f_0+m\Delta f\right)\left(\xi-\hat{\xi}\right)\frac{m+1/2}{M}\frac{\tau}{\xi}\right] \frac{\sin\left\{\pi\left(f_0+m\Delta f\right)\left(\xi-\hat{\xi}\right)\frac{1}{M}\frac{\tau}{\xi}\right\}}{\pi\left(f_0+m\Delta f\right)\left(\xi-\hat{\xi}\right)\frac{1}{M}\frac{\tau}{\xi}}$$

$$= \frac{\sigma_s(z)}{M}\frac{\tau}{\xi} \sum_{m=0}^{M-1} \exp\left[j\phi\right]\frac{\sin\upsilon}{\upsilon}$$

where

$$\phi = 2\pi(f_0+m\Delta f)(\xi-\hat{\xi})\frac{m+1/2}{M}\frac{\tau}{\xi}$$

and

$$\upsilon = \pi(f_0+m\Delta f)(\xi-\hat{\xi})\frac{1}{M}\frac{\tau}{\xi}.$$

Somewhat surprisingly, the resolution of this method for estimating Doppler shift ξ is no better than using the same pulse length unmodulated. This is shown in Figure 4.45.

Recently, StratoSonde has attempted to develop an FM-CW SODAR system in which the frequency linearly changes with time, with a very long (typically 40 s) pulse duration. The configuration is strictly bistatic, although the transmitter and receivers are quite close physically. This concept is fraught with technical problems, including: difficulties with the direct signal from the transmitter beating with the echo signals ("beats" are the lower frequencies heard when two sounds of nearly the same frequency are added); problems associated with the atmospheric scattering being of differing strength at different wavelengths; the Bragg condition not being

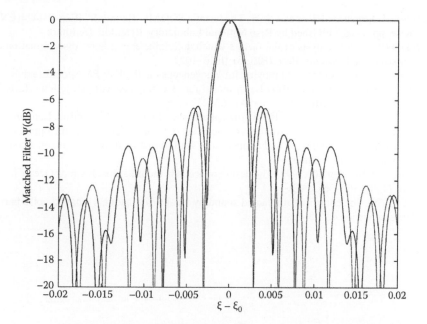

FIGURE 4.45 Matched filter output for $M = 5$ frequency steps, $f_T = 3400$ Hz, and pulse duration $T = 0.1$ s (solid curve). Also shown is the equivalent output using an unmodulated pulse of the same duration (dotted curve).

met because of a range of wavelengths in the scattering volume; and beating between signals reflected from different heights. These problems have been at least partially solved, but result in a very expensive, computationally intensive, and bulky instrument which may not have any substantial resolution advantages over conventional unmodulated systems.

4.14 SUMMARY

In this chapter the ideas underlying the scattering of sound and reception of a signal have been developed. This led to the acoustic radar equation, which provides insights into how the instrument design impacts on the signal quality. This aspect is explored further by detailed investigation of speakers, microphones, antenna design, baffles, and signal modulation. The information in this chapter really defines the SODAR.

REFERENCES

Bradley SG (1999) Use of coded waveforms for SODAR systems. Met Atmos Phys 71(1–2): 15–23.

Georges TM, Clifford SF (1972) Acoustic sounding in a refracting atmosphere. J Acoust Soc Am 52(5): 1397–1405.

Georges TM, Clifford SF (1974) Estimating refractive effects in acoustic sounding. J Acoust Soc Am 55(5): 934–936.

Girardin-Gondeau J, Baudin F et al. (1991) Comparison of coded waveforms for an airborne meteorological Doppler radar. J Atmos Ocean Technol 8: 234–246.

Mikkelsen T, Jørgensen HE et al. (2005) The bistatic SODAR "Heimdall" Risø-R-1424(EN), Risø, pp. 1–28. published by Risø National Laboratory, Roshilde, Denmark

North DO (1963) An analysis of the factors which determine signal noise discrimination in pulsed carrier systems. Proc IEEE 51: 1016–1027.

Ostashev VE (1997) Acoustics in moving inhomogeneous media. E & FN Spon, London.

Phillips PD, Richner H et al. (1977) Layer model for assessing acoustic refraction effects in acoustic echo sounding. J Ac Soc Am 62. 277–285.

Rogers CD (2000) Inverse methods fro atmospheric sounding. World Scientific, London.

Schomburg A, Englich D (1998) Analysis of the effect of acoustic refraction on Doppler measurements caused by wind and temperature. 9th International Symposium on Acoustic Remote Sensing, Vienna.

Werkhoven CJ, Bradley SG (1997) The design of acoustic radar baffles. J Atmos Ocean Technol 14(3I): 360–367.<AQ>

Wirt LS (1979) The control of diffracted sound by means of thnadners (shaped noise barriers). Acoustica 42: 73–88.

5 SODAR Systems and Signal Quality

In Chapter 4, the theoretical basis was given for transmission of sound in directed beams and receiving echo signals. The basis for interpreting these signals in terms of turbulent parameters and wind speed components was discussed in detail. In particular, it is evident that acoustic beam patterns are seldom simple and that interpretation of echo signals requires knowledge of the remote-sensing instrument design. In this chapter we discuss the details of actual designs, so the connection can be made between hardware elements in Chapter 5 and the theoretical considerations of Chapter 4.

5.1 TRANSDUCER AND ANTENNA COMBINATIONS

5.1.1 SPEAKERS AND MICROPHONES

Speakers are generally piezoelectric horn tweeters for higher frequency phased-array systems (such as the Motorola KSN1005 or equivalent used in the AeroViron-ment 4000) or high-efficiency coil horn speakers (such as the RCF 125/T similar to that used in the Metek SODAR/RASS) for lower frequency phased-array systems, or high-power cone drivers (such as the Altec Lansing 290-16L) for single-speaker dish systems (Fig. 5.1).

Speakers are specified as having sensitivity of a particular intensity level L_I generally measured at a distance of 1 m for 1 W input electrical power

$$L_I = 10 \log_{10} \left(\frac{I}{10^{-12} \, \mathrm{W \, m^{-2}}} \right) \tag{5.1}$$

for acoustic intensity I.

For example, the KSN1005 has an output of 94 dB for 2.83 V_{rms} input voltage, measured at a distance of 1 m, or 2.5 mW m^{-2}. The 2.83 V_{rms} reference gives $2.83^2/8$ = 1 W into 8 Ω, which is a common speaker resistance value. Since the conversion to acoustic power is an electrically lossy process, equivalent to a resistance, power output is proportional to V_{rms}^2, and also the intensity is inversely proportional to the square of the distance, so the intensity produced at distance z is

$$I_z = 0.0025 \left(\frac{V_{rms}}{2.83z} \right)^2 = 0.3 z^{-2} \, \mathrm{mW \, m^{-2}}.$$

The maximum allowable input is 35 V_{rms}, giving 3.1 W m^{-2} at 1 m distance. The frequency response for this tweeter is shown in Figure 5.2. The 3-dB point below the quoted 97 dB is at 4 kHz.

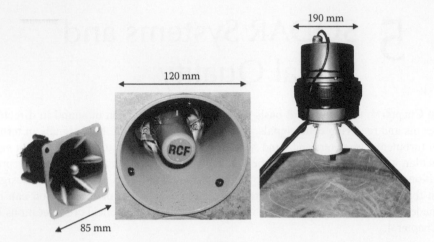

FIGURE 5.1 Some speakers used in research SODARs. From left to right: Motorola KSN1005, RCF 125/T, and Altec lansing 290-16L.

For the purposes of modeling performance, a good fit to the angular patterns in Figure 5.3 is obtained using $I_{max} \cos^4 \theta$, with $I_{max} = 0.31$, 1.9, 3.1, and 3.1 W m^{-2} for 35 V$_{rms}$ at 1 m and for frequencies $f_T = 3.15$, 4, 5, and 6 kHz. Integrating over the forward hemisphere

$$2\pi I_{max} \int_0^{\pi/2} \cos^4 \theta \sin \theta \, d\theta = \frac{2\pi}{5} I_{max} = 0.39, 2.4, 3.9, \text{and } 3.9 \text{ W}$$

for the total *acoustic* power. Measurements show that this speaker's impedance at f = 4 kHz is about 250 Ω, and is equivalent to a 0.12 μF capacitance in parallel with

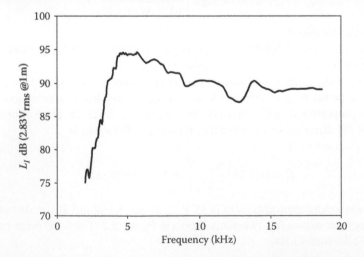

FIGURE 5.2 The measured frequency response of a Motorola KSN1005A speaker.

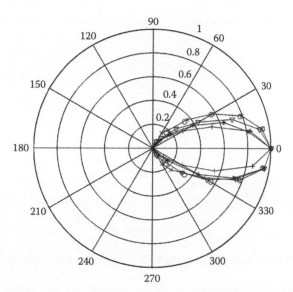

FIGURE 5.3 Polar patterns of normalized intensity for the KSN1005 speaker at 3.15 kHz (x), 4 kHz (dotted line), 5 kHz (*), 6 kHz (+), and $\cos^4 \theta$ (circles).

a 1 kΩ resistor (Figure 5.4). This means that the electrical power dissipated from 35 V_{rms} input is 1.2 W. The electric-acoustic power conversion efficiency is therefore around 50% at 1 kHz. For monostatic use, this speaker is used as a microphone. Its sensitivity was measured in comparison with a calibrated microphone, giving the points in Figure 5.5.

Similar measurements can be performed on other speakers. The RCF 125/T is quoted as having a 750 Hz cutoff and 120 dB re 1 V/1 m: its diameter is 120 mm. The 290-16L has 3 dB cutoff at 300 Hz and a speaker diameter of 190 mm (but horn diameter of 90 mm).

Note that the diameter of the speaker is related to its low-frequency 3 dB cutoff frequency, as shown in Figure 5.6 for these three speakers.

Some speaker specifications also quote their sensitivity as a microphone. For example, the Four-Jay 440-8 has an output of 108 dB at 2 kHz for 1 W electrical input into the 8 Ω, and a receiver sensitivity of 13.7 mV_{rms} output for 1 Pa (i.e. L_I = 94 dB) input. Note that sensitivity of coil speakers is generally much less than for piezoelectric speakers. These figures can be compared with, for example, the Knowles MR8540 microphone which has a sensitivity of 6.3 μV for 1 Pa input.

From the combination of acoustic power output as a speaker and voltage input as a microphone, it is possible to calculate the overall system gain $V_{microphone}/V_{speaker}$ for a single speaker or for an

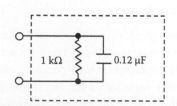

FIGURE 5.4 The equivalent electrical circuit for a KSN1005 speaker.

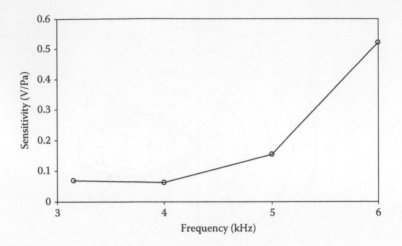

FIGURE 5.5 Measured sensitivity of the KSN1005 used as a microphone.

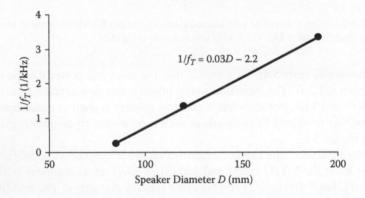

FIGURE 5.6 The upper frequency 3 dB point for three speakers versus their diameter.

array. For example, with the KSN1005, 3×10^{-4} Wm^{-2} is obtained at 1 m for 1 V$_{rms}$ input, corresponding to $20 \times 10^{-6}(3 \times 10^{-4}/10^{-12}) = 0.35$ Pa. A KSN1005 placed at 1 m will record $0.1 \times 0.35 = 0.035$ V$_{rms}$ output. With a Four-Jay 440-8, $10^{10.8-12}/8 = 7.9 \times 10^{-3}$ W m^{-2} or 1.8 Pa, giving 0.024 V$_{rms}$ output at an identical Four-Jay 440-8 at 1 m.

5.1.2 HORNS

All the speakers mentioned above have an acoustic horn connecting the driver element to the atmosphere. The horn acts as an impedance-matching element from the small-displacement high-pressure speaker diaphragm to a large-displacement lower-pressure variation in the air. Horns generally have the diaphragm area larger than the throat area: the ratio is called the compression ratio of the horn. For midrange frequency the compression ratio is typically 2:1, and high-frequency tweeters can have compression ratios as high as 10:1.

Information on horn design can readily be found in texts or web pages, but a rough guide is that the length of the horn should be about the longest wavelength, λ_L, which is going to be used, and the mouth of the horn should have a circumference equal to or greater than λ_L. So for a 2-kHz system, the horn would be about 170-mm long and 54-mm diameter. Horns generally have an exponential flare, rather than being conical, but for higher frequencies the shorter tractrix shape is common:

$$x = \ln\left(1 + \sqrt{1 - r_x^2}\right) - \ln(r_x) - \sqrt{1 - r_x^2},$$

where x = (distance from the mouth)/(radius of the mouth) and r_x =(radius at distance x)/(radius of mouth) – in other words dimensions are scaled by the mouth radius which is typically $\lambda_L/2\pi$.

The beam pattern from a horn having a mouth radius a is again just the pattern from a hole of radius a,

$$P \propto \left[2 \frac{J_1\left(ka \sin\theta\right)}{ka \sin\theta}\right]^2.$$

5.1.3 PHASED-ARRAY FREQUENCY RANGE

The beam polar pattern is the product of the speaker polar pattern and the array or dish pattern. The individual speaker pattern changes with frequency: Figure 5.7 shows the measured pattern for a single KSN1005 at 4 and 6 kHz. It is clear that the array pattern will dominate over the small changes in the individual speaker pattern.

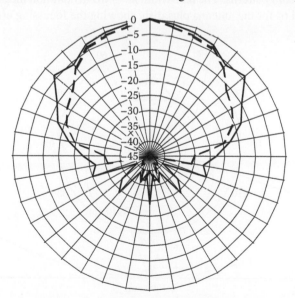

FIGURE 5.7 Polar patterns for an individual KSN1005 at 4 kHz (solid line) and 6 kHz (dashed line).

The first minimum from an array consisting of $M \times M$ speakers separated by distance d is given by Eq. (4.8) as $\Delta\theta \approx c/Mdf_T$, so for reasonably large arrays the beam width is inversely proportional to frequency. A more narrow and intense beam is desirable. Eq. (4.3), giving the first two side lobe zenith angles θ_L on either side of the main beam, can be expressed in the form

$$\sin\theta_L = -\frac{3c}{4f_T d}, \frac{5c}{4f_T d}$$

(5.2)

if an incremental phase shift of $\pi/2$ is used. If the next main lobe is kept below a zenith angle of $45°$, $3c/(4f_T d) > 1/\sqrt{2}$. If beams are directed at $45°$ to rows or columns of close-packed speakers, then d can be replaced by $d/\sqrt{2}$. A useful guide based on the second lobe position and the relationship between speaker efficiency and its diameter (in m) is therefore $c/(2f_T d) > 1$ and $f_T > 1000/(30d - 2.2)$, or

$$\frac{1000}{30d - 2.2}\frac{d}{c} < \frac{f_T d}{c} < \frac{3}{2}$$

(5.3)

For example, for the KSN1005, this gives $3\ kHz < f_T < 6\ kHz$. Extensive field tests with the AeroVironment 4000 have proven these to be practical limits.

5.1.4 DISH DESIGN

As an example of a dish antenna design, Figure 5.8 shows a 3-beam system based on the Four-Jay 440-8 re-entrant cone speaker and a 1.2-m dish. Figure 5.9 shows the measured beam patterns. The half-width at −3 dB (a common measure) is $25°$ for the speaker and $6°$ for the antenna plus dish, showing the focussing effect described

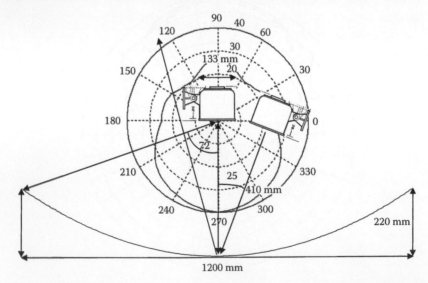

FIGURE 5.8 The design of a dish-based 3-beam system.

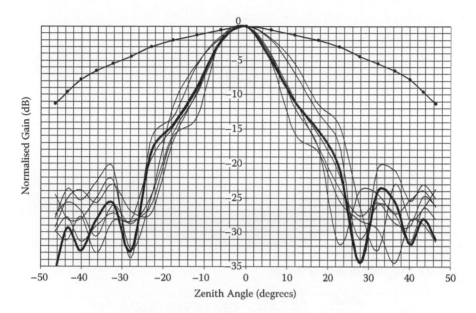

FIGURE 5.9 Measured beam patterns for the dish system at 3 kHz: speaker pattern without dish (line with dots); speaker at calculated focal distance (solid line); speaker at other positions within ±50 mm of the calculated focus.

earlier. Note that diffraction effects can easily be seen past about 25° for the dish system. Figure 5.10 shows a spun aluminum dish. In this prototype, the distance of the speaker from the dish can be adjusted, since the equivalent source point within the speaker horn is not known.

5.1.5 DESIGNING FOR ABSORPTION
AND BACKGROUND NOISE

Obviously absorption is lower at lower frequencies. The absorption is of order $0.003f_{kHz}^2$ dB m^{-1} at 50% relative humidity and 10°C. Roughly speaking, the difference between $f_T = 2$ and 6 kHz is an extra 10 dB lost per 100 m. This is a lot.

From Chapter 3, background noise decreases roughly as f_T^{-q}, so higher transmitting frequencies are favored. But since background noise depends on a *power* of f_T and absorption depends on the *exponential* of frequency-dependent absorption times range, there will be an optimum frequency for any given range. The ratio of received signal power to received acoustic noise power (SNR) is written as

$$\text{SNR} = A \frac{f_T^{1/3} \exp(-2 \propto z)}{f_T^{-q}} = A f_T^{q+1/3} \exp(-2bf_T^2 z), \qquad (5.4)$$

so

FIGURE 5.10   A dish antenna system.

$$\frac{\text{dSNR}}{\text{d}f_T} = A\left[\frac{q+1/3}{f_T} - 4bzf_T\right]f_T^q \exp(-2bf_T^2 z)$$

and the optimal f_T for a fixed range z is

$$f_T = \sqrt{\frac{q+1/3}{4bz}}$$

(5.5)

The slope of the background noise spectrum for the daytime city is about $q = 2.8$ so for a range of $z = 1000$ m, given $b = 0.003/10 \log_{10}e = 7 \times 10^{-4}$ m^{-1}, the optimum $f_T = 1$ kHz. In practice this is a little pessimistic, since good signal processing can extend the optimum frequency by about a factor of 2, as shown in Figure 5.11.

5.1.6   REJECTING RAIN CLUTTER

Scattering from rain depends on f_T^4, so lower frequencies give markedly less spectral noise from rain. For example, the SNR in rain will be around 20 dB better at $f_T = 2$ kHz than at 4.5 kHz: high-frequency mini-SODARs have real problems during rain! However, acoustic noise from drop splashing is likely to be greater at lower frequencies.

Figure 5.12 shows measurements taken on five different roofing panel structures (Hopkins, 2004). These comprise: 25-mm thick polycarbonate sheet (five layers of 3.4 kg m^{-2}); laminated glazing (6-mm toughened glass, 12-mm air space, 6.4-mm laminate glass); and ETFE pillows of a 150-micron layer taped to a 50-micron layer with a 200-mm air gap with and without two types of rain suppressors. The rain noise in all cases decreases as $f^{-3/2}$. This means that the overall effect of rain, considered as a noise source, varies as $f_T^{5/2}$, so that lower frequency SODARs perform better.

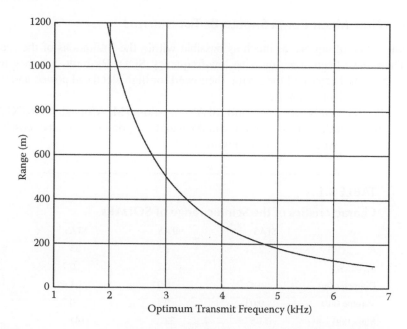

FIGURE 5.11 The optimum transmit frequency for a given range, determined by the balance between decreasing background noise and increasing absorption with increasing frequency.

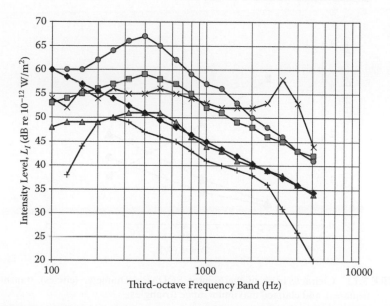

FIGURE 5.12 Spectral intensity levels measured on ETFE (circles), polycarbonate (x), ETFE with rain suppressor type 1 (squares), ETFE with rain suppressor type 2 (triangles), and laminated glazing (+). Also shown is a curve having an $f^{-3/2}$ dependence (black diamonds).

5.1.7   How Much Power Should Be Transmitted?

The answer is, of course, as much as possible within the limitations of the speakers. There have been some massive low-frequency SODARs built, but they have little popularity because of their bulk, their need for high electrical power, and their obtrusive environmental noise.

The Scintec combination of small (SFAS), medium (MFAS), and large (XFAS) phased-array SODARs uses similar technology and is a good indication of cost/benefit versus power (see Table 5.1 and Figure 5.13).

TABLE 5.1

Characteristics of the Scintec range of SODARs

	SFAS	MFAS	XFAS
$P_{acoustic}$ (W)	2.5	7.5	35
$P_{12\,v}$ (kW)	0.1	0.2	0.7
Diameter (m)	0.42	0.72	1.45
Volume (m³)	0.03	0.1	0.7
Mass (kg)	11.5	32	144
f_T (kHz)	3.2	2.2	1.0
z_{min} (m)	10	20	20
z_{max} (km)	0.5	1	2

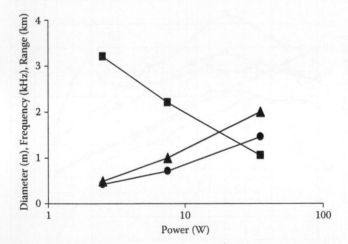

FIGURE 5.13 Characteristics of the Scintec SODARs. Diameter (circles), transmit frequency f_T (squares), and claimed maximum range (triangles).

5.2   SODAR TIMING

5.2.1   PULSE SHAPE, DURATION, AND REPETITION

SODARs generate a pulse which has the generic shape shown in Figure 5.14. The key parameters are transmit frequency f_T, pulse period τ, and ramp up/down time $\beta\tau$. Transmission of such pulses is repeated with pulse repetition rate T as shown in Figure 5.15.

Because there are multiple beams in a monostatic system, the pulse repetition rate for an individual beam will be the number of beams times the repetition rate for transmitting. The power transmitted is proportional to the pulse length τ, for a given pulse amplitude. Also, the Doppler spectrum frequency resolution is better with a longer pulse. This can be visualized by estimating f_T by counting the number of cycles, n, in time τ, and then

$$f_T = \frac{n}{\tau}$$

If there is a ±1 uncertainty in n, then the uncertainty in f_T is

$$\Delta f_T = \pm\frac{1}{\tau}$$

(5.6)

The spectral line from a constant Doppler shift is therefore spread to $2/\tau$ wide. The practical Doppler resolution is actually better than this because of averaging and peak-detection schemes, as discussed later, but the spectral width is still a basic limitation.

However, a longer transmitted pulse means a longer range gate and poorer spatial resolution. Basically two layers cannot be distinguished if they are within vertical distance $c\tau/2$ of each other.

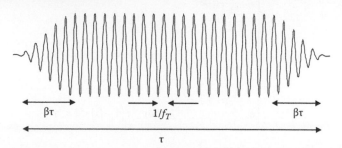

FIGURE 5.14   The generic shape of SODAR pulses of duration τ.

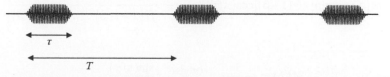

FIGURE 5.15   The repetition of pulses in a SODAR system.

A practical compromise seems to be $\tau \sim 30$ to 80 ms, giving $\Delta f_T \sim 10$ to 30 Hz, or (raw) uncertainty in horizontal wind speed of 1 to 3 m/s, and spatial resolution of 5 to 14 m.

The pulse repetition rate, T, is determined by the highest range z_T from which echoes are expected. It is important that this is chosen conservatively (i.e., pick a much larger T than the range of interest), otherwise echo returns from higher than this range, from an earlier pulse, will add to those from lower down due to the current pulse. This means that echo returns are combined from heights $ct/2$ and $c(t+T)/2$. For example, the AeroVironment 4000 typically has $T = 1.33$ s, giving $z_T = 340 \times 1.33/2 = 220$ m, for the 200 m typically analyzed.

It is desirable to shape the start and end of the pulse as shown, since this reduces oscillations in the frequency spectrum, and consequently limits spreading of power from a spectral peak into adjacent spectrum bins. To do this, the pulse voltage is typically multiplied by a Hanning shape

$$m(t) = \begin{cases} \dfrac{1}{2}\left[1 - \cos\left(\dfrac{\pi}{\beta\tau}t\right)\right] & \text{for } 0 < t < \beta\tau \\[3mm] 1 & \text{for } \beta\tau < t < \tau(1-\beta) \\[3mm] \dfrac{1}{2}\left[1 - \cos\left(\dfrac{\pi}{\beta\tau}\{\tau - t\}\right)\right] & \text{for } \tau(1-\beta) < t < \tau \end{cases}$$

(5.7)

or a Gaussian shape

$$m(t) = \begin{cases} e^{-\frac{1}{2\sigma_m^2}(t-\beta\tau)^2} & \text{for } 0 < t < \beta\tau \\[3mm] 1 & \text{for } \beta\tau < t < \tau(1-\beta) \\[3mm] e^{-\frac{1}{2\sigma_m^2}(t-\{1-\beta\}\tau)^2} & \text{for } \tau(1-\beta) < t < \tau \end{cases}$$

(5.8)

These have Fourier transforms of

$$M(f) = \tau \sin\left[\pi f\tau(1-\beta)\right]\cos\left(\pi f\tau\beta\right)\left[\frac{1}{\pi f\tau} - \frac{1}{\pi f\tau(2+1/\beta)} - \frac{1}{\pi f\tau(2-1/\beta)}\right]$$

(5.9)

and

$$M(f) = \frac{1}{\sqrt{2\pi}\sigma_f}e^{-(1/2\sigma_f^2)f^2} * \frac{\sin(\pi f\tau)}{\pi f\tau},$$

(5.10)

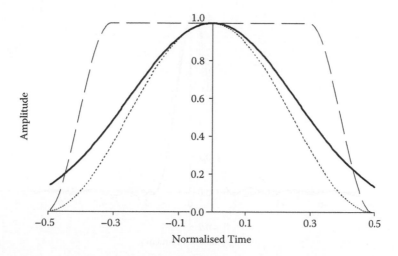

FIGURE 5.16 Pulse envelopes for a Gaussian with $\sigma_m = \tau/4$ (dark solid line), Hanning with $\beta = 0.2$ (light line), and Hanning with $\beta = 0.5$ (dashed line).

where $*$ means the convolution product and $\sigma_f = 1/(2\pi\sigma_m)$. Pulse envelopes are shown in Figure 5.16 and their corresponding spectra in Figure 5.17.

It is clear from Figure 5.17 that a smoother pulse envelope produces a smoother and wider spectrum. The smoothness is desirable, since it reduces the possibility of secondary maximum adding to noise to give a spurious Doppler peak estimate and hence a false wind estimate. On the other hand, a wider spectrum makes it more difficult to estimate the point of highest curvature (the spectral peak position). In all cases, the spectral shape can be estimated by a Gaussian of width σ_f in the central region. For the Hanning case,

$$\sigma_f \tau = 0.52 + 0.73\beta \tag{5.11}$$

and for the Gaussian case with $\sigma_m = \tau/4$,

$$\sigma_f \tau = 0.64. \quad . \tag{5.12}$$

5.2.2 RANGE GATES

The received signal depends on the convolution of the atmospheric turbulent scattering cross-section profile with the pulse envelope, as expressed in (4.33). For the zero-Doppler case,

$$p_R(t) \propto \int_0^\infty \sigma_s(z) m(t - t_z) \exp[j2\pi f_T(t - t_z)] \mathrm{d}z.$$

The pulse shape $m(t)$ and duration τ determine spatial resolution of the SODAR through this term. *Spatial resolution* is the vertical separation Δz_m of two infinitely

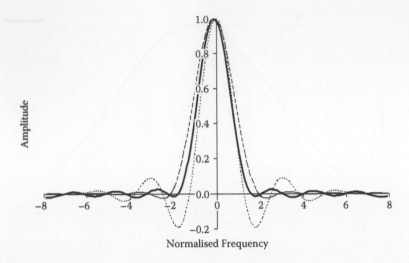

FIGURE 5.17 Frequency spectra corresponding to a Gaussian envelope with $\sigma_m = \tau/4$ (dark solid line), a Hanning envelope with $\beta = 0.2$ (light line), and a Hanning envelope with $\beta = 0.5$ (dashed line).

thin layers which is resolved in the returned signal. Two peaks in the time series are resolved if the signal drops to at least half power between them. If σ_s consists of delta functions at heights z_1 and z_2, then

$$p_R(t) \propto \int\limits_0^\infty [\delta(z-z_1)+\delta(z-z_2)]m\left(t-\frac{2z}{c}\right)\exp\left(-j2\pi f_T\,\frac{2z}{c}\right)dz$$

$$\propto m\left(t-\frac{2z_1}{c}\right)\exp\left(-j2\pi f_T\,\frac{2z_1}{c}\right)+m\left(t-\frac{2z_2}{c}\right)\exp\left(-j2\pi f_T\,\frac{2z_2}{c}\right),$$

which has an envelope of

$$m\left(t-\frac{2z_1}{c}\right)+m\left(t-\frac{2z_2}{c}\right).$$

For the Gaussian case, if $z_2 = z_1 + \Delta z_m$, then the minimum of the combined envelope pattern occurs at $t = (2z_1/c + 2z_2/c)/2$ or $t-2z_1/c = \Delta z_m/c$ and $t-2z_2/c = -\Delta z_m/c$. The situation is shown in Figure 5.18. At this time, the peaks are resolved if

$$2\exp\left[-\frac{1}{2\sigma_m^2}\left(\frac{\Delta z_m}{c}\right)^2\right] \le \frac{1}{\sqrt{2}}$$

or

$$\Delta z_m \ge c\sigma_m\sqrt{\ln 8}. \tag{5.13}$$

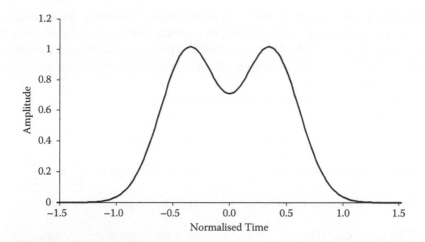

FIGURE 5.18 The combination of two Gaussian envelopes which allows spatial features to just be resolved.

For $\sigma_m = \tau/4$, $\Delta z_m > 0.36c\tau$. For a square pulse envelope, $\Delta z_m > 0.5c\tau$ for two spatial features to be resolved.

This spatial resolution for turbulence measurements is determined by the pulse shape. In practice, the SODAR will usually sample much more rapidly than this, but this does not increase spatial resolution.

More importantly for many applications is the spatial scale resolved for wind vectors. Wind components are estimated from the Doppler shift in the peak power in a power spectrum. Each power spectrum is obtained from a Fourier transform of a set of M data values sampled at time intervals of Δt. This means that wind component estimates are only obtained from height intervals of

$$\Delta z_V = \frac{c(M\Delta t)}{c} \tag{5.14}$$

Again, SODARs will often present results at finer spatial resolution, perhaps by doing fast Fourier transforms (FFTs) using overlapping sequences of samples. While this may look good on a profile plot, no extra information is contained.

For example, assume that a 2 kHz SODAR has pulse length $\tau = 50$ ms, and the atmosphere has $c = 340$ m s^{-1}, a constant σ_s, and Doppler shift of -45 Hz below $z_0 = 85$ m and $+45$ Hz above that level. The recorded time series consists of a pure sine wave at 1960 Hz for the first 0.5 s, a mixture of 1960 and 2040 Hz until 0.55 s, and then a pure tone at 2040 Hz. The signal is sampled at $f_s = 960$ Hz for $M = 64$ points, producing samples at frequencies 960/64 = 15 Hz apart.

The spatial resolution due to the FFT length is $\Delta z_V = 11.3$ m and that due to pulse length is $\Delta z_m = 8.5$ m. Spectral resolution due to the finite sampling length of $T = M/f_s = 67$ ms is $1/T = 15$ Hz (the first zeros of the spectrum are at 45±15 Hz). If the finite pulse length is included, the spectral resolution is now $1/\tau = 20$ Hz. In fact the pure tone spectral line is convolved with *both* the spectrum from the finite sampling length and the spectrum from the finite pulse length. Convolving the spectrum is equivalent to multiplying the time series by a rectangular function. In this case the time series is being multiplied by *two* rectangular functions, and this is equivalent to simply multiplying by the shorter rectangle. So the spectral resolution is determined by the *shorter* of T and τ.

To summarize spatial and spectral resolutions:

Spatial resolution for turbulence: $\Delta z_m = \dfrac{c\tau}{2}$,

Spatial resolution for winds: $\Delta z_V =$ the larger of $\dfrac{c\tau}{2}$ and $\dfrac{cM}{2f_s}$,

Wind speed spectral resolution: $\Delta f_V =$ the larger of $\dfrac{f_s}{M}$ and $\dfrac{1}{\tau}$,

Wind speed resolution: $\Delta V =$ the larger of $\dfrac{cf_s}{2Mf_T}$ and $\dfrac{c}{2\tau f_T}$

Uncertainty product for winds: $\Delta z_V \Delta V = \begin{cases} \dfrac{c^2}{4f_T}\left(\dfrac{f_s\tau}{M}\right)^{-1} & \text{if } \dfrac{f_s\tau}{M} < 1, \\[4mm] \dfrac{c^2}{4f_T}\left(\dfrac{f_s\tau}{M}\right) & \text{if } 1 < \dfrac{f_s\tau}{M}. \end{cases}$

From the above it is clear that the minimum of the resolution product is when $\tau = M/f_s = T$. Then

$$\Delta z_V \Delta V = \frac{c^2}{4f_T}$$

(5.15)

Typically, $c = 340$ m s^{-1} and $f_T = 4000$ Hz, so $\Delta z_V \Delta V \approx 7$ m^2 s^{-1}. If $\Delta z_V = 10$ m, then $\Delta V = 0.7$ m s^{-1}. This is a good first guide, but later it will be seen that good peak detection and averaging can improve velocity resolution substantially.

5.3   BASIC HARDWARE UNITS

5.3.1   THE BASIC COMPONENTS OF A SODAR RECEIVER

All SODAR receivers consist of some common components: microphones to convert acoustic power into electrical power; amplifiers to provide large enough voltages for digital processing; filters to reject unwanted noise; and digitization modules.

5.3.2   MICROPHONE ARRAY

Most SODARs are monostatic, so use the speaker as a microphone. This precludes using a sensitive microphone. Horn speakers are generally used, where the small speaker driver is impedance matched to the atmosphere via a horn-shaped extension.

A typical phased array made of 64 of the 0.085-m square KSN1005 speakers will have an area of $64 \times 0.085 \times 0.085 = 0.46$ m^2 and an equivalent radius of $a \sim 0.4$ m.

The power received from turbulence at 100 m is of order $10^{-14}GA_e$, or $\sim 10^{-14}$ W, giving an intensity of $10^{-14}/0.46 \sim 2 \times 10^{-14}$ W m^{-2} at 1 m. Normal microphone sensitivities vary from about -20 dB referred to 1 V/Pa (or 0.1 V/Pa) for a carbon microphone, to -90 dB re 1 V/Pa for a ribbon microphone. Sound pressure is approximately $30\sqrt{I}$ Pa for intensity I in Pa, or about 4 µPa from the turbulence. This means that normal microphones will give from about 10^{-12} to 10^{-7} V$_{rms}$ output.

The voltage produced by the piezoelectric KSN1005, acting as a microphone, should be around 10^{-8} V$_{rms}$, or 0.6 µV$_{rms}$ for the whole array.

Moving coil speakers are also commonly used for lower-frequency SODARs. Moving coil microphones are typically two orders of magnitude less sensitive than piezomicrophones, but the atmospheric absorption coefficient is almost an order of magnitude smaller for a 1.6 kHz system compared to a 4.5 kHz system. The result is perhaps an order of magnitude smaller signal, say 60 nV$_{rms}$.

Note that with such small signals, some care is necessary with electrical shielding and grounding.

5.3.3 Low-Noise Amplifiers

Typical outputs from the speaker/microphone array are 100 to 1000 nV$_{rms}$, so around 120 dB voltage gain (10^6) is required in the receiver to produce signals in the vicinity of 1 V for digitization. In practice, the microphone/speaker self-noise and other external acoustic noise will generally be larger (meaning some signal averaging will be needed), but a good design goal would be to minimize that component of the noise over which the designer has control. The equivalent RMS noise voltage in a 100 Hz bandwidth at the input of a good low-noise operational amplifier is 10 nV, about 10% of the expected input signal, so it is important to choose the preamplifier carefully and to take care with circuit layout and ground connections. It is also important to keep input resistance small, so that resistor noise does not contribute significantly.

As an example, a common low-noise op-amp is the AD OP-27E, having 3 nV Hz$^{-1/2}$ noise at its input. For 100 Hz bandwidth this gives 30 nV noise at the input. A gain of 1000 (60 dB) can readily be used with this op-amp, using say 10Ω input resistors and 10 kΩ feedback resistors, as shown in Figure 5.19.

Resistor noise can be reduced further using parallel resistors, since the resistor noise in each resistor is uncorrelated, whereas the input currents from the desired

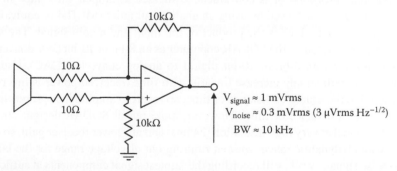

FIGURE 5.19 A typical preamplifier input stage.

signals are correlated. For example, if s_1 and s_2 are two signals with the same signal mean and same (uncorrelated) noise levels:

$$\overline{s_1 + s_2} = \overline{s_1} + \overline{s_2} = 2\overline{s}$$

$$\sigma^2_{s_1 + s_2} = \frac{1}{T} \int_0^T \left[(s_1 + s_2) - \overline{s_1 + s_2} \right]^2 dt$$

$$= \frac{1}{T} \int_0^T \left[(s_1 - \overline{s_1}) + (s_2 - \overline{s_2}) \right]^2 dt$$

$$= \frac{1}{T} \int_0^T (s_1 - \overline{s_1})^2 dt + \frac{2}{T} \int_0^T (s_1 - \overline{s_1})(s_2 - \overline{s_2}) dt + \frac{1}{T} \int_0^T (s_2 - \overline{s_2})^2 dt$$

$$= \frac{1}{T} \int_0^T (s_1 - \overline{s_1})^2 dt + \frac{1}{T} \int_0^T (s_2 - \overline{s_2})^2 dt$$

$$= \sigma^2_{s_1} + \sigma^2_{s_2} = 2\sigma^2_s$$

$$SNR = \frac{2\overline{s}}{\sqrt{2\sigma^2_s}} = \frac{1}{\sqrt{2}} \frac{\overline{s}}{\sigma_s}$$

(5.16)

so the SNR decreases by $1/M^{1/2}$ for M signals added together. This is a particularly useful technique for phased arrays consisting of many speakers/microphones. For example, an array of 64 microphones will have an SNR improvement of a factor of 8 in amplitude or 18 dB in power. Some filtering can also be usefully done at this point, by including capacitors across each of the two 10 kΩ resistors.

5.3.4 RAMP GAIN

Since the echo signal decreases with distance (and therefore time) due to beam spreading and absorption, it is convenient to include a ramped gain stage in the receiver. This can be achieved by using an analog multiplier (MLT04 or equivalent) which has an output which is the product of the input and a gain signal. The gain signal can be generated by the SODAR computer as an 8-bit or 10-bit code converted to analog form via an 8-bit or 10-bit digital to analog convertor (DAC). Usually the gain signal will simply increase linearly with time (received power is inversely proportional to the square of distance or time, so received amplitude is proportional to the inverse of distance or time). However, more recent SODAR designs simply digitize the signal at very high resolution (24 bits) and at a lower receiver gain, so that there is enough dynamic range, without running out of voltage range for the larger signal+noise signals, while still recording the faintest signal components at sufficient resolution. In these designs, *all* processes such as filtering and ramp gain are done in software, as well as allowance for absorption, depending on measured temperature and/or humidity.

5.3.5 FILTERS

Random electronic noise can easily be 30% of the signal received from 100 m. Hardware filters can be used to improve this SNR. Generally a relatively simple band-pass filter might follow the preamplifier. The bandwidth (BW) required is the maximum Doppler shift

$$\Delta f = \frac{2V_{\max} \sin \varphi}{c} f_T. \tag{5.17}$$

Typically the maximum wind speed capability is 25 m s^{-1} (at which speed wind noise is often significant), and the beam tilt angle is ~20°, so the required Q of the BP filter is

$$Q = \frac{f_T}{\Delta f} \approx 20 \tag{5.18}$$

(the Q factor is a measure of a filter's selectivity). Typically a 4-pole pair BP filter with Q = 10 to 20 would be used at this stage, and might have a gain of 10 (i.e., 20 dB). This could be a unit purchased as a complete module, or comprise an active filter IC and some tuning components, or be built up from op-amps. It could also be a digitally programmable filter if it were desired to be able to change f_T. Programmable filters can be based on ICs such as the LTC1068 which requires a tuning input at 100 or 200 times the desired center frequency. Modular programmable BP filters are also available, such as the Frequency Devices 828BP which has an 8-bit parallel programmable center frequency. Typical values are given in the circuit of Figure 5.20, with a voltage transfer function shown in Figure 5.21 for a 2-kHz SODAR. The output from the filter will have noise reduced by a factor of 10 in comparison with the signal.

5.3.6 MIXING TO LOWER FREQUENCIES (DEMODULATION)

In practice, all the useful information is contained in the signal amplitude and in the Doppler shift, so any pure signal component at frequency f_T can be removed. The

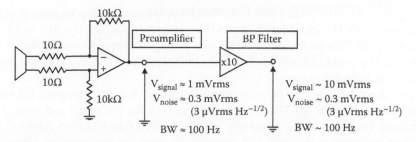

FIGURE 5.20 The effect of a band-pass filter stage.

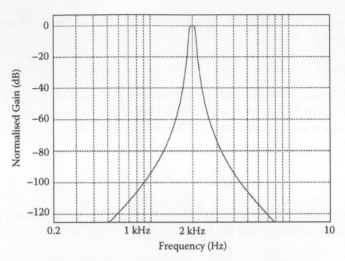

FIGURE 5.21 A typical voltage transfer function for a band-pass filter.

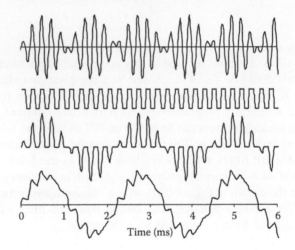

FIGURE 5.22 Demodulation of a modulated signal (upper trace) by multiplying with a mixing signal (second trace from the top) to give a composite low-frequency and $2f_T$ signal (third trace) which can be low-pass filtered to give the modulation (bottom trace).

mixing process can be understood from the plot in Figure 5.22 of a modulated echo signal (top trace). The second trace shows the mixing signal which is multiplied with the echo signal. This produces the third trace. Finally, a simple low-pass filter, such as provided by an RC circuit, produces the smoothed bottom trace. This last trace contains the modulation signal.

This demodulation can be accomplished with an analog multiplier IC, such as the Analog Devices MLT04, or by switching between the signal and ground at the mixing frequency using an analog switch IC. The mixer is then followed by an LP filter to remove the higher frequencies, as shown in the complete circuit of Figure 5.23.

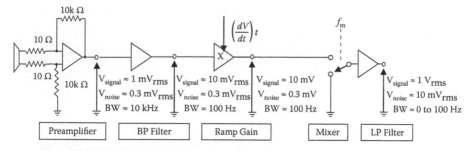

| Preamplifier | BP Filter | Ramp Gain | | Mixer | LP Filter |

FIGURE 5.23 The complete amplifier and filter chain.

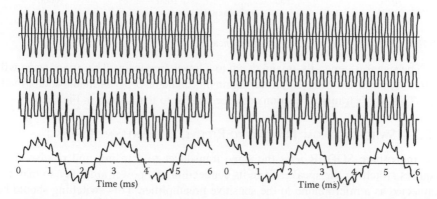

FIGURE 5.24 Demodulation of a Doppler-shifted (FM) signal. Mixing with a square wave in phase with the original transmitted signal is shown on the left, and mixing with a quadrature phase (90° phase-shifted) square wave on the right.

It is seen that the initial power SNR of $20 \log_{10}(1 \text{ mV}/0.3 \text{ mV}) = 10$ dB has been increased to $20 \log_{10}(1 \text{ V}/0.01 \text{ V}) = 40$ dB.

Similarly, FM modulation produced by Doppler shift will be demodulated using this mixer, as shown in Figure 5.24. Of course, the modulation is in practice of much lower frequency than the transmitted frequency, and a sharp cutoff LP filter gives a much smoother output than shown in the figure.

For this example, it can be seen that the mixing frequency is 4.5 kHz. From the FM demodulated (low frequency) traces, the period of the Doppler component is seen to be about 2.2 ms ($\Delta f = 450$ Hz). The in-phase mixed demodulated signal lags the 90°-phase demodulated signal by 90°. This is a case of *positive* Doppler shift, with the raw signal frequency being 4.95 kHz.

If, on the other hand, the raw signal frequency is 4.05 kHz, the in-phase trace is the same as in Figure 5.24, but the 90°-phase trace is inverted, and the in-phase mixed demodulated signal leads the 90°-phase demodulated signal by 90°. This is a case of *negative* Doppler shift. So it can be seen that the relative phase of the in-phase and 90°-phase demodulated signals shows whether the wind component is toward the SODAR (positive shift) or away from the SODAR (negative shift).

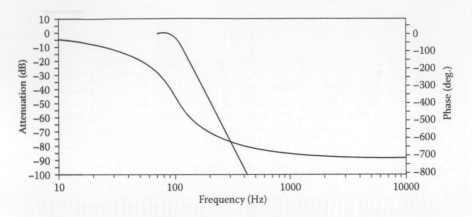

FIGURE 5.25 The gain and phase of a typical LP filter.

The signal is generally mixed down to a lower frequency. The mixer stage will be followed by a low-pass (LP) filter. Again this could be programmable and/or modular. This filter's transfer function could be similar to that shown in Figure 5.25.

5.3.7 Switching from Transmit to Receive, and Antenna Ringing

The monostatic SODAR uses the same transducer to transmit and receive. This requires switching the speaker from its connection to a power amplifier so that it is connected as a microphone to the sensitive preamplifier. This switching should be done as rapidly as possible after the end of the transmitted pulse, so that echoes from low altitudes can be analyzed.

There are a number of problems associated with this switching. First, the switch must be an analog switch (i.e., allow continuously varying signals to pass through it). Secondly, it must handle relatively high voltages and currents during transmitting (of order 100 V into 16 Ω for a coil speaker, giving 6 A), as well as the very small voltages and currents during receiving (of order 1 μV into 10 Ω, giving 100 nA). Switching should be stabilized after the equivalent of a few meters of pulse travel (say 20 ms). In addition, there must be very good isolation of the preamplifier from the power amplifier, and care must be taken that transients do not destroy the preamplifier.

In spite of these difficulties, a simple relay such as the Omron G2RL has a current rating of 8 A, a turn-on time of 7 ms, and a turn-off time of 2 ms, and will be sufficient. More sophisticated semiconductor switches (TRIACs, etc.) can also be used.

The real problem with recording useful data at a low altitude is that the antenna and the baffle enclosure are likely to "ring" for some time after the transmit pulse. This is not simply the time taken for sound to travel along the baffle and back to the speakers, since a typical speed in a composite wooden baffle might be 10^3 m s^{-1} and for a length of 2 m this would only give a return time of 4 ms. The problem is reverberation time of both the baffles and the speaker enclosure. A good design will attempt to damp any reverberations. This can be approached by using "soft" materials for the baffle, such as composite wood, perhaps coated with a matting or lead layer, and by filling the speaker enclosure with acoustic foam and perhaps other "deadening" materials. Even so, the problems with reverberations are likely to affect

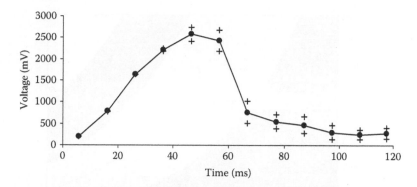

FIGURE 5.26 Signal amplitude levels recorded from an AeroVironment 4000 SODAR within the first 120 ms following the transmit pulse.

data quality for at least the lowest 6 to 10 m (40 ms). Figure 5.26 shows a typical transient from an AeroVironment 4000 SODAR.

The SODAR settings for this example were 100% amplitude and 60-ms pulse length. Generally it is to be expected that more power and longer pulse lengths will increase reverberation. Protecting the preamplifier input from transients, such as reverberations, is simply a matter of installing protecting diodes across all resistors and the input of the preamplifier. Genuine echo signals will always be sufficiently small that these diodes will not be turned on.

5.4   DATA AVAILABILITY

5.4.1   THE HIGHEST USEFUL RANGE

A 2-kHz SODAR might range to, say, 400 m, and a typical 4.5 kHz system might range to 200 m at a quiet country daytime site (of course depending on turbulence intensities). At these heights the SNR will be around 1. Most of the *difference* in range capability will be due to absorption dependence on frequency. The absorption coefficients are, for 50% relative humidity and at 10°C, $\alpha \approx 0.01$ dB/m (0.002 m^{-1}) at 2 kHz, and $\alpha \approx 0.05$ dB/m (0.01 m^{-1}) at 4.5 kHz. From Chapter 3, the frequency- and range-dependent terms in the SNR are, from the SODAR equation,

$$f^{q+\frac{1}{3}}\frac{e^{-2\alpha z}}{z^2}$$

(5.19)

and the ratio of these for the two frequencies and ranges is of order 1. However, as seen in Chapter 3, the noise dependence on frequency, q, is about 2.8 for city backgrounds, 1.4 for daytime country, and 0.5 for nighttime country. Combining these concepts allows an estimate of SNR versus SODAR frequency, depending on site background noise. This is plotted in Figure 5.27, assuming constant backscatter with height, independent of site and time of day. The turbulence levels vary substantially, but, for example, if a $z^{-4/3}$ dependence for C_T^2 is assumed (as found by various groups for convective boundary layers) then the SNR would fall off more rapidly with height.

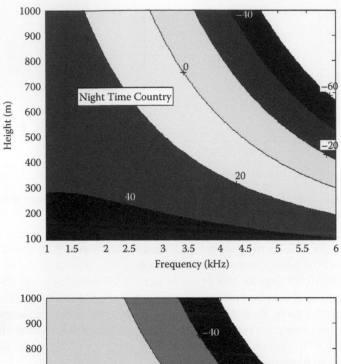

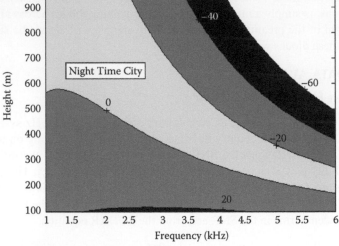

FIGURE 5.27 SNR versus height and frequency for night time country environments (upper plot), night-time city environments (middle plot), and daytime city environments (lower plot).

5.5 LOSS OF SIGNAL IN NOISE

One of the principal problems of ground-based remote sensing is the poorer data availability at greater heights, and the fact that data availability depends on meteorological conditions. The SODAR equations can be written as

$$P = P_A + P_F + P_P + P_N, \qquad (5.20)$$

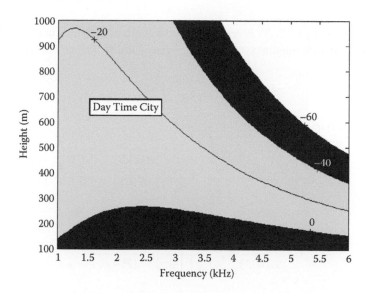

FIGURE 5.27   Lower plot.

where P is the total received power, P_A is the power scattered from atmospheric turbulence, P_F the power reflected from fixed objects such as masts, P_P the power scattered from precipitation, and P_N noise power. The required signal is from P_A and the remaining terms on the right lead to reduced SNR = $P_A/(P_F+P_P+P_N)$.

Generally P_F may be reduced by selecting the orientation of the SODAR to minimize power transmitted toward the fixed object. If P_F is still present, then it can often be identified because it has zero Doppler shift and its spectral width may be different from that of P_A. While fixed echoes remain an *operational problem*, for calibration purposes and even in many data collection applications, those range gates affected can simply be ignored.

Echoes from precipitation are also an operational problem for SODARs, but these data can effectively be eliminated because the presence of rainfall can be sensed via other means or from the increased vertical velocities detected by the SODAR.

External noise remains the main difficulty during calibration. Both P_A and P_N can be variable. From the SODAR equation

$$P_A = [3\times10^{-4}\, P_T GA_e \tau f_T^{1/3}]\left[\frac{c^{2/3}}{T^2}\right]\left[\frac{e^{-2\alpha z}}{z^2}\right]C_T^2. \tag{5.21}$$

The first square bracket contains factors determined by the instrument, and the second square bracket contains terms only weakly dependent on atmospheric temperature profile variations. The third square bracket contains terms representing signal loss due to absorption and spherical spreading, and the C_T^2 term represents the echo signal generation. The absorption is generally not very large, so most signal loss is through the unavoidable inverse-square reduction with height. For example, the inverse square loss between 10 and 100 m is 20 dB, whereas the absorption loss is around 0.6 dB for a 1 kHz SODAR and 6 dB for a 4.5 kHz SODAR.

As seen in Chapter 2, C_T^2 is related to the strength of turbulence, which depends on both site (surface roughness) and atmospheric stability. From (2.16) and (2.18)

$$C_T^2 = \frac{0.106}{0.033} \varepsilon^{-\frac{1}{3}} \varepsilon_\Theta$$

(5.22)

From (2.19) and (2.20)

$$\varepsilon = K_h \left(\frac{\overline{du}}{dz}\right)^2 (P_r - R_i) \quad \text{and} \quad \varepsilon_\Theta = K_h \left(\frac{\overline{d\Theta}}{dz}\right)^2 = K_h \left(\frac{\overline{T}}{g}\right)^2 R_{(i)}^2 \left(\frac{\overline{du}}{dz}\right)^4$$

giving

$$C_T^2 = \frac{0.106}{0.033} K_h^{2/3} \left(\frac{\overline{T}}{g}\right)^2 \left(\frac{\overline{du}}{dz}\right)^{10/3} \frac{R_i^2}{(P_r - R_i)^{1/3}}.$$

(5.23)

Data availability is determined by C_T^2/z^2, so if the wind shear is largely determined by the site, the variations in C_T^2 and data availability are largely determined by R_i. Figure 5.28 shows two contours of constant C_T^2/z^2 superimposed on the data availability diagram Bradley et al. (2004). Near neutral conditions, where $R_i = 0$, this theory appears to hold, but for larger absolute values of R_i there seems to be

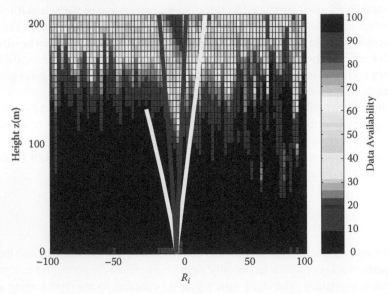

FIGURE 5.28 (See color insert following page 10). Percentage of relative data yield of Scintec SODAR receptions, plotted against height z of the SODAR range gates and against the Richardson number R_i based on meteorological mast measurements at 100 m. The solid yellow and blue lines are two contours of constant C_T^2/z^2..

little dependence on R_i. Instead, above about 100 m, availability is roughly proportional to $1/z$ rather than $\sqrt{C_T^2}/z$. This is shown in Figure 5.29. Similarly, availability is shown as a function of z/L in Figures 5.30 and 5.31 for L computed at two different heights. While Figure 5.27 and Figure 5.30 are qualitatively similar, as predicted, the 20 m L gives quite different patterns (Figure 5.31), possibly because this is in a different scaling and turbulence regime.

The above shows that systematic bias can occur in calibration and operationally due to reduced data availability with extended height and during near-neutral conditions. In this section we have attempted to find a measurable and commonly used parameter (either R_i or L) which can be used in a functional relationship of the form

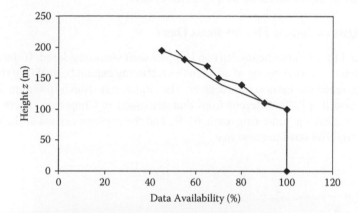

FIGURE 5.29 Data availability based on Figure 5.28 and on $1/z$ for heights above 100 m.

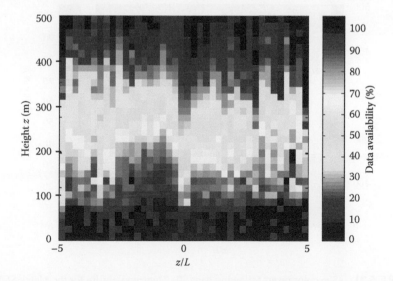

FIGURE 5.30 (See color insert following page 10). Data availability for the Metek SODAR based on Monin-Obhukov length L estimated from a sonic anemometer at 20 m height.

Data availability = f(R_i, z) or Data availability = g(L,z).

If successful, this approach would allow inversion of the function to give a prediction of availability under different atmospheric conditions. On the basis of Figure 5.28, Figure 5.29, and Figures 5.30 and 5.31 neither R_i nor L give a complete description and further work clearly needs to be done. The problem is that we essentially are attempting to predict C_T^2 from R_i or L, since availability should be closely correlated with echo strength, which is directly proportional to C_T^2. Ideally, perhaps, C_T^2 would be measured independent of the SODAR, and used to predict the SODAR data availability, but this approach would require purpose-designed mast installations. Another approach would be to use a complex boundary layer model to predict C_T^2, but this would be an entire new research effort.

5.5.1 Loss of Signal Due to Beam Drift

In Chapter 4 the effect of beam drift on Doppler shift was considered. If the acoustic beam is blown sideways by the wind, then the scattering cannot be at 180° if the return signal is to reach the monostatic receiver. The situation is shown in Figure 5.32.

This case is a little different from that discussed in Chapter 4. There are two important angles, the scattering angle $\theta_r - \theta_s$, and the angle of arrival at the receiver, θ_r. It is straightforward to show that

$$\theta_s = \theta_r - \theta_s = \tan^{-1}\frac{V}{c}. \tag{5.24}$$

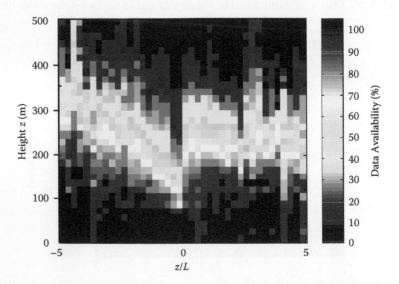

FIGURE 5.31 (See color insert following page 10). Data availability for the Metek SODAR based on Monin-Obhukov length L estimated from a sonic anemometer at 100 m height.

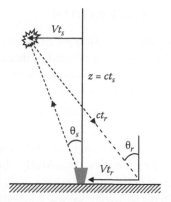

FIGURE 5.32 The geometry of beam drift for the vertical beam of a monostatic SODAR.

There are two competing effects: a reduction in signal level due to the off-axis avrrival; and an increase in signal level due to the inclusion of a component of C_V^2. The combined fractional change in signal level can be written in the form

$$v = \left[2\frac{J_1(ka\sin\theta_r)}{ka\sin\theta_r}\right]^2\left[1+7.33\left(\frac{T}{c}\right)^2\cos^2\left(\frac{\theta_r-\theta_s}{2}\right)\frac{C_V^2}{C_T^2}\right]$$

$$\approx \left\{2\frac{J_1[ka\sin(2\theta_s)]}{ka\sin(2\theta_s)}\right\}^2\left[1+5\cos^2\left(\frac{\theta_s}{2}\right)\frac{C_V^2}{C_T^2}\right].$$

(5.25)

This function is plotted in Figure 5.33 with $T^2C_V^2 = 100C^2C_T^2$.

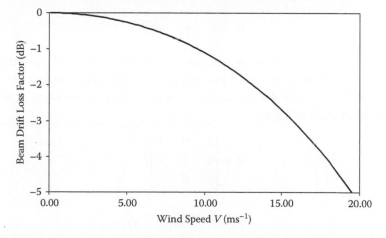

FIGURE 5.33 The loss of signal due to the beam drifting with the wind, assuming $T^2C_V^2 = 100C^2C_T^2$, c = 340 m s⁻¹, and ka = 20.

5.6   CALIBRATION

Most calibration effort for SODARs is directed toward obtaining good-quality wind profiles. In the case of wind estimates, factors such as unknown absorption amounts, and unknown power output do not come into play. The main factor is the Doppler frequency shift which, in principle, is easy to measure.

It is possible to calibrate a SODAR for C_T^2 and C_V^2 using sonic anemometers mounted on a mast. Although C_T^2 and C_V^2 are conceived as scale-independent quantities, differences might arise if two instruments sensed these quantities at very different length scales. Thankfully, the scale of most sonic anemometers is not very different from the $\lambda_T/2$ scale sensed by a SODAR.

In the following we will concentrate on calibration of a SODAR for wind estimation. Some of the considerations will also be applicable to calibrations for turbulent structure function parameters.

5.6.1   WHY ARE CALIBRATIONS REQUIRED?

In general, a number of beams are transmitted and received. Each of these produces a Doppler spectrum from which peak detection processes obtain a Doppler shift Δf corresponding to the radial component of velocity for that beam. The scaled Doppler shift, $\eta = -\dfrac{c}{2} \dfrac{\Delta f}{f_T}$, for each beam, is a linear sum of contributions from velocity components u, v, and w. The system of equations can be written

$$N = DV = BTRV \tag{5.26}$$

where

$$N = (\eta_1 \ \ \eta_2 \ \ \eta_3 \ \ \eta_4 \ \ \eta_5)^{\mathrm{T}} \tag{5.27}$$

$$V = (u \quad v \quad w)^{\mathrm{T}} \tag{5.28}$$

and the superscript T indicates the transpose of a matrix. D is the Doppler-transformation matrix composed of a beam pointing factor B, an orientation factor (rotation of the instrument around the z axis) R, and an out-of-level or tilt factor T. Ideal or expected values of T and R are

$$\hat{T} = \begin{pmatrix} 1 & 0 & 0 \\ 0 & 1 & 0 \\ 0 & 0 & 1 \end{pmatrix}, \tag{5.29}$$

$$\hat{R} = \begin{pmatrix} \cos\phi & \sin\phi & 0 \\ -\sin\phi & \cos\phi & 0 \\ 0 & 0 & 1 \end{pmatrix}. \tag{5.30}$$

For a 5-beam monostatic SODAR having four tilted beams at the same zenith angle, and one vertical beam, the expected values of B are

$$\hat{B} = \begin{pmatrix} \sin\theta & 0 & \cos\theta \\ 0 & \sin\theta & \cos\theta \\ 0 & 0 & 1 \\ -\sin\theta & 0 & \cos\theta \\ 0 & -\sin\theta & \cos\theta \end{pmatrix}$$

(5.31)

and for a 3-beam monostatic SODAR having two tilted beams at the same zenith angle, and one vertical beam,

$$\hat{B} = \begin{pmatrix} \sin\theta & 0 & \cos\theta \\ 0 & \sin\theta & \cos\theta \\ 0 & 0 & 1 \end{pmatrix}.$$

(5.32)

For a 5-beam bistatic SODAR having four receiver beams at the same zenith angle and one vertical beam,

$$\hat{B} = \begin{pmatrix} \sin\theta & 0 & -1-\cos\theta \\ 0 & \sin\theta & -1-\cos\theta \\ 0 & 0 & -2 \\ -\sin\theta & 0 & -1-\cos\theta \\ 0 & -\sin\theta & -1-\cos\theta \end{pmatrix}$$

(5.33)

and for a 3-beam bistatic SODAR having two receiver beams at the same zenith angle and one vertical beam,

$$\hat{B} = \begin{pmatrix} \sin\theta & 0 & -1-\cos\theta \\ 0 & \sin\theta & -1-\cos\theta \\ 0 & 0 & -2 \end{pmatrix}.$$

(5.34)

If the *actual* B, T, and R matrices are the same as the *expected* matrices for a particular instrument, then (5.26) is readily inverted to find the wind components V from the measurements N. In such a case, no calibration is required.

The actual beam matrix may differ because of some change in configuration or because of not completely understanding the beam shape. For example, with the 3-beam monostatic, if all beams are in reality effectively tilted $\Delta\theta$ more than expected,

$$B = \begin{pmatrix} \sin[\theta+\Delta\theta] & 0 & \cos[\theta+\Delta\theta] \\ 0 & \sin[\theta+\Delta\theta] & \cos[\theta+\Delta\theta] \\ 0 & 0 & 1 \end{pmatrix}.$$

(5.35)

Similarly, the instrument could be an angle α out of level, for example, with rotation around the x-axis, giving

$$
T = \begin{pmatrix} 1 & 0 & 0 \\ 0 & \cos\alpha & -\sin\alpha \\ 0 & \sin\alpha & \cos\alpha \end{pmatrix}.
$$

$$(5.36)$$

The result is that the expected velocity matrix is

$$
\hat{V} = (\hat{D}^T\hat{D})^{-1}\hat{D}^T N = [(\hat{D}^T\hat{D})^{-1}\hat{D}^T D]V = CV, \tag{5.37}
$$

where C is the calibration matrix. This formulation is the least-squares solution (Rogers, 2000) for velocities, and allows for more beams than the three velocity components. For the three-beam case, a simpler (but equivalent) form is

$$
\hat{V} = \hat{D}^{-1}N = \hat{D}^{-1}DV = CV. \tag{5.38}
$$

Ideally $C = I$, the identity matrix.

The effect of a leveling error is quite different from the effect of a tilt-angle error. For example, for a 3-beam monostatic system having $\theta = 20°$ and $\phi = 0$, and with $u=w=0$, a leveling error $\alpha = 1°$ gives $m = 0.9998$, whereas a tilt error of $\Delta\theta = 1°$ gives $m = 1.0478$. It is very important therefore to have good calibration of the effective beam direction. The 5-beam monostatic system gives the same result:

$$
C = \begin{pmatrix} 1.0478 & 0 & 0 \\ 0 & 1.0478 & 0 \\ 0 & 0 & 0.9949 \end{pmatrix}.
$$

The calibration operation is carried out to find C. In principle, it is then possible to decompose C into its known part $\hat{D}$ and unknown part D, and subsequently analyze the changes in D which do not agree with expectations. In practice, it seems this is rarely done, with researchers and manufacturers being content just to find C.

Most calibrations concentrate on obtaining correlations between horizontal wind speeds and directions. In (5.37) all three wind components from V must be kept, as well as the full forms for C, and only after $\hat{V}$ is found can the u and v components be squared and added. This is because an out-of-level error can cause some contamination of horizontal velocity components from w. As a simple example, consider a 3-beam monostatic SODAR having $T = R = I$, but with an error in beam tilt angle given by (5.35). Then $D = B$ and

$$C = \begin{pmatrix} \dfrac{\sin(\theta+\Delta\theta)}{\sin\theta} & 0 & \dfrac{\cos(\theta+\Delta\theta)-\cos\theta}{\sin} \\[2ex] 0 & \dfrac{\sin(\theta+\Delta\theta)}{\sin\theta} & \dfrac{\cos(\theta+\Delta\theta)-\cos\theta}{\sin\theta} \\[2ex] 0 & 0 & 1 \end{pmatrix}$$

(5.39)

and

$$\begin{pmatrix} \hat{u} \\ \hat{v} \\ \hat{w} \end{pmatrix} = \begin{pmatrix} \dfrac{\sin(\theta+\Delta\theta)}{\sin}u + \dfrac{\cos(\theta+\Delta\theta)-\cos\theta}{\sin}w \\[2ex] \dfrac{\sin(\theta+\Delta\theta)}{\sin\theta}v + \dfrac{\cos(\theta+\Delta\theta)-\cos\theta}{\sin\theta}w \\[2ex] w \end{pmatrix}.$$

(5.40)

Then

$$\hat{u}^2 + \hat{v}^2 = \left[\frac{\sin(\theta+\Delta\theta)}{\sin\theta}u + \frac{\cos(\theta+\Delta\theta)-\cos\theta}{\sin\theta}w\right]^2 + \left[\frac{\sin(\theta+\Delta\theta)}{\sin\theta}v + \frac{\cos(\theta+\Delta\theta)-\cos\theta}{\sin\theta}w\right]^2.$$

Given that w is small compared to u or v and $\Delta\theta \ll \pi$,

$$\hat{u}^2 + \hat{v}^2 \approx \left(1 + 2\frac{\Delta\theta}{\tan\theta}\right)(u^2 + v^2) = m^2(u^2 + v^2).$$

(5.41)

Here m is the calibration slope for a fit through the origin. This solution is close to, but not the same as that given by,

$$\hat{V}^T\hat{V} = (CV)^TCV = V^T(C^TC)V.$$

5.6.2 Effective Beam Angle

The main beam lobe for either the Bessel function or the sin function form can readily be fit (very accurately) with a Gaussian. For example, the shape of the angular variation in power from a dish antenna of radius a, with it axis pointing at angle θ_0, is well represented over the central region by a Gaussian:

$$\left[\frac{2J_1\left(ka\sin\{\theta-\theta_0\}\right)}{ka\sin(\theta-\theta_0)}\right]^2 \approx 1 - \frac{(ka)^2(\theta-\theta_0)^2}{4} \approx \exp\left[-\frac{(ka)^2(\theta-\theta_0)^2}{4}\right].$$

Since a volume element at radial distance r is $r^2dr\sin\theta d\theta d\phi$, the volume/power-weighted radial wind component multiplier is

$$P \propto \exp\left[-\frac{(ka)^2(\theta-\theta_0)^2}{4}\right]\sin^2\theta.$$

This peaks when

$$\frac{1}{P}\frac{dP}{d\theta} = \frac{2}{\tan\theta} - \frac{(ka)^2(\theta-\theta_0)}{2} = 0$$

or

$$\Delta\theta = \theta - \theta_0 \approx \frac{4}{(ka)^2 \tan\theta_0}.$$

In other words, there is a systematic bias to the wind speed estimates by a factor of

$$m-1 \approx \frac{4}{(ka)^2 \tan^2\theta_0}.$$

(5.42)

For example, if $a = 0.4$ m, $k = 60$ m^{-1}, and $\theta_0 = 0.3$, then $m-1 = 0.073$, a 7.3% error in calibration. It is likely that a failure to properly account for volume/power weighted Doppler shift leads to the main source of wind speed error in SODARs.

5.6.3 What Accuracy Is Required?

The growth of wind turbine heights has introduced a requirement for high-accuracy wind profiling. In order that SODARs can be used instead of cup anemometers on masts, the accuracy obtained from the calibration should be comparable to the accuracy of the cups. Sufficient accuracy is probably better than 1% of wind speed which translates roughly into 1% in direction. The pressure force on a turbine blade of area A is, from Bernouli's equation, $\rho V^2 A/2$, and the power transferred is the product of force and speed, or $\rho V^3 A/2$. This means that power is proportional to the cube of wind speed and an accuracy of 1% in wind speed gives an accuracy of 3% in wind power estimation.

5.6.4 Calibrations against Various Potential Standards

Atmospheric Research and Technology Ltd. produces SODAR self-calibration tools (http://www.sodar.com/prod02.htm). These, however, check the basic functionality of a SODAR rather than conducting a calibration. For example, the frequency transmitted is checked and the ability of the SODAR to detect peaks, but there is no check on the beam pattern and directivity nor is there any check on actual response to winds.

A group at Salford University is developing a transponder which essentially detects the sound generated by the SODAR over a wide solid angle and produces a delayed, Doppler-shifted, response which has noise-added based on real atmospheric echoes and on the SODAR equation. Such a transponder would require distributed microphones and speakers mounted above the SODAR on some kind of thin framework and in the far field (perhaps 10-m distance). This would check most of the functions of the SODAR and the error conditions described, but leveling and calibrating this device would also be critical since it would be extremely undesirable to simply transfer a calibration problem from one device to another.

Wind speed and direction have often been measured by free flying wind-sonde balloons, with RADAR reflectors or geophysical satellites] and compared with the SODAR output. The difficulty is that the balloon profiles are very transitory compared to a typical wind profile averaging time of 5 to 10 minutes for a SODAR. The nominal ascent rate of a Vaisala sonde is 4 m s^{-1}, so the balloon will return data from the lowest 200 m for 50 s: this corresponds to about seven complete cycles around the beams of a 5-beam SODAR. During these 50 s, the balloon will have also drifted 500 m downwind if the wind speed is 10 m s^{-1}. These problems mean that free balloons are never going to provide adequate calibration for SODARs.

Some tethered balloon systems also record winds. For example, the Vaisala DigiCora tethersonde system (http://www.vaisala.com/) has up to six wind sensors on the tether (say with spacing 30 m for a 200-m line) with accuracy of 0.1 m s^{-1} and 1° in direction. In practice, though, swinging of the sensors might be expected to give much reduced accuracy in the field. Because they also require a large amount of manpower for operation, they can only be used intermittently. Therefore, systems of this sort are not intended to be primary standards.

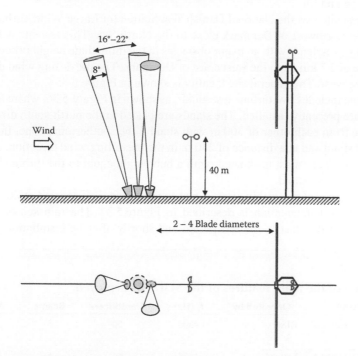

FIGURE 5.34 Preferred calibration configuration. A SODAR is situated in the prevailing upwind direction from the turbine and at a distance of two to four diameters. The acoustic beams are aimed away from the turbine and mast to minimize fixed echoes.

The expense of routine calibration against cup anemometers on a tall ($\geq$100 m) mast precludes this method except as verification that a short-mast calibration method is satisfactory and that there is not some calibration problem unaccounted for. This was the focus of the PIE field trials described below (Bradley et al., 2004). In the following, we will use these relatively recently conducted trials as an example of field calibration methodology and outcomes.

Many of the calibration problems referred to above can be resolved by calibrating the SODAR against a cup anemometer and/or sonic mounted on a short (40 m, say) mast. The geometric errors described above can be eliminated and the estimation bias largely reduced, leaving only the difficulties arising if the SNR is inadequate. The preferred situation, for the wind turbine power performance context, is shown in Figure 5.34.

5.6.5   THE PIE FIELD CAMPAIGN SETUP

The Profiler Inter-comparison Experiment (PIE) was conducted in order to test the calibration of a range of SODARs against a well-instrumented 120-m mast (Antoniou et al., 2004, 2005). The SODARS were operated at sufficiently different frequencies to not interfere with each other. The SODARs deployed are described in Table 5.2. The AV4000 had a Model 3000 enclosure, which may have modified the beam shape and tilt.

The test site was the National Danish Test Station for Large Wind Turbines situated in the northwest of Denmark close to the North Sea. This test site is flat, surrounded by grassland, with no major obstacles in the immediate neighborhood and at a distance of 1.7 km from the west coast of Denmark. The prevailing wind direction is from the west. The general site locality is shown in Figure 5.35.

The site includes five turbine test stands, as shown in Figure 5.36, where five wind turbines are presently installed. The stands are placed in the north-south direction at a distance from each other of 300 m with stand 5 the southernmost one. In front of every test stand and at a distance of 240 m in the prevailing wind direction, a meteorological (or 'met') mast is situated, with a hub height equal to the turbine height at the corresponding stand.

At the south of the turbine row, a met tower (stand 6) is located at 200 m from stand 5. Its instrumentation is described in Figure 5.37. The rain sensor was not installed from the start of the test period and shortly after its installation it failed.

TABLE 5.2

Summary of the main features of the SODARs deployed

SODAR	Operated by	f_T (Hz)	Transducers	Beams	Δz (m)
AeroVironment 4000	RISØ	4500	50	3	10
Metek PCS2000-64 with 1290 MHz RASS	Salford	1674	64	5	15
Scintec SFAS	WINDTEST KWK	2540–4850	64	9	5

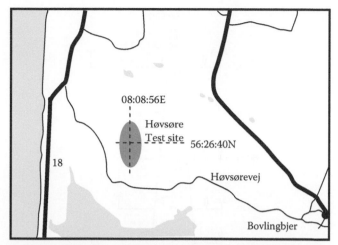

FIGURE 5.35   The PIE locality, with the site indicated by the dashed oval.

FIGURE 5.36   The test site, the met tower, and the SODARs (wind blowing from the east).

Later on a tipping bucket rain sensor was installed. Likewise, the 100 m wind direction sensor was not available from the beginning of the measurement period due to a lightning strike. Four phased-array SODARs were located to the southwest of the meteorological (or 'met') tower (Figure 5.38), but the high-frequency Metek SODAR was not used for calibrations.

5.6.6   Raw SODAR Data versus Mast

Initial plots of raw SODAR wind speed versus cup wind speed show a number of problems, typified by Figure 5.39.

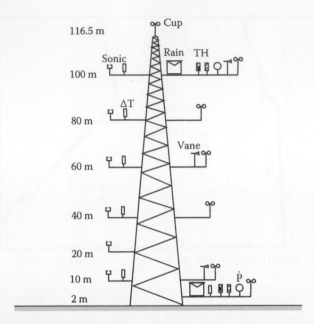

FIGURE 5.37 Schematic representation of the tower instrumentation (looking to the tower from the west). Instruments consist of cup anemometers ("cup"), wind vanes ("vane"), sonic anemometers ("sonic"), differential temperature transducers ("ΔT"), absolute temperature transducer "T", humidity sensors "H", and pressure sensors "P".

FIGURE 5.38 The three SODARs used for calibration. From upper left: Scintec SFAS (octagonal baffle), AV4000 (small square baffle), and Metek SODAR/RASS (large square baffle). The fourth SODAR, lowermost in the picture, was not used for calibration.

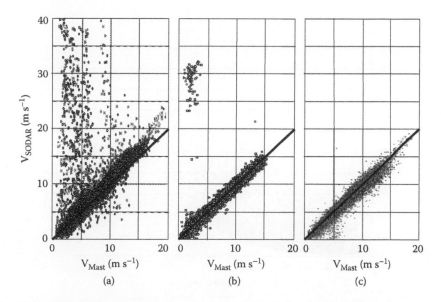

FIGURE 5.39 Mast wind speeds at 40, 60, 80, 100, and 116m versus raw SODAR wind speeds. (a) AV4000, (b) Scintec, and (c) Metek.

The raw plots have obvious outlier data points. For the AV4000 and the Scintec, high apparent SODAR winds at low mast winds are caused by rain. The particular filtering options selected for the various SODARs in these examples remove most of these points for the Scintec and virtually all rain points for the Metek, but other choices could give opposite results.

A second cause of outliers is fixed echoes. These points exhibit low apparent winds from the SODAR and higher winds from the cup anemometers, as evidenced by the points in Figure 5.39(c).

5.6.7 NUMERICAL FILTERING OF DATA

The SODAR data first must be filtered to remove rain data, fixed echo data, and any other bad data which are due to external noise.

The mast anemometers are known to have a sector from which winds do not give good data because of shielding by the mast. In the case of the PIE trial, the wind direction sector from 325–90° gave potentially contaminated cup data. The Scintec SODAR–Mast data set was filtered to remove these data, but much of the data shown for the other two SODARs includes all wind directions (however, see the detailed analysis below for the Metek SODAR).

During rain, the signal is backscattered from falling raindrops, resulting in a relatively large negative vertical velocity. This velocity contaminates the horizontal wind calculations and can lead to predictions of high wind speeds. Rain gauges were part of the mast instrumentation, but it is also possible to filter the SODAR data based on just SODAR observations to remove most of the rain contamination. This is a desirable approach, since it removes the need for yet another instrument when

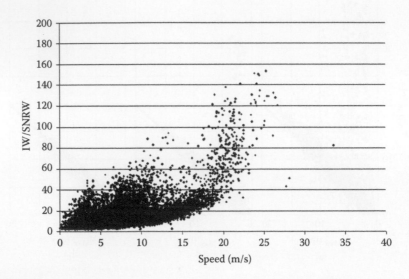

FIGURE 5.40 Vertical beam intensity IW divided by vertical beam SNR versus wind speed during dry periods.

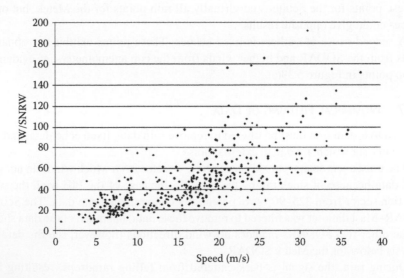

FIGURE 5.41 Vertical beam intensity divided by SNR versus wind speed during raining periods.

SODARs are used as autonomous wind sensors at wind energy installations. Each SODAR has methods for detection of "bad data" and in many cases the menu-guided user choices can allow for automatic removal of rain-contaminated data. Additionally, or independently, it is possible to use the routinely available diagnostic information provided by the SODAR output to construct a rain-rejection filter.

For example, the AV4000 outputs a quantity called IW, the intensity of the echo from the vertical (w) beam, and a corresponding SNR for that beam called SNRW. During rain IW will generally be higher and SNRW lower, so the ratio IW/SNRW is a possible rain discriminator. On the AV4000 dry periods typically show IW/SNRW = 20 to 50, whereas in rain IW/SNRW may rise to about 100. During snowfall there is some evidence that IW/SNRW is *lower* than the dry figure. A test was run at the ECN EWTW site, using a tipping bucket rain gauge for comparison. Figure 5.40 shows IW/SNRW versus wind speed during dry periods, and Figure 5.41 for raining periods. It can be seen from these figures that IW/SNRW is only indicative of the *possibility* of rain occurring. This analysis suggests that, without further research, rain gauges should be deployed to indicate possible contamination of wind data by rainfall.

Echoes from hard, static, non-atmospheric objects ("fixed echoes") result in a peak at zero frequency shift which, depending on atmospheric signal strength, can be wrongly interpreted as a radial wind speed of 0 m s⁻¹. SODAR manufacturers generally have some fixed echo detection and removal filters which can be applied with some success.

There are two steps required to filter the data completely:

1. The parameters of the SODAR manufacturer's software need to be chosen so that most of the faulty data points are filtered out before any manual data analysis is undertaken. There is no general rule as to which parameters to choose as these vary between different SODARs and sites. The parameters depend on the digital signal processing that is used for data acquisition. For that reason it has so far been impossible to agree on a common filter standard for all SODAR manufacturers and models.
2. In a second step the data set has to be evaluated manually to include filters that depend on external measurement parameters such as the sector filtering and the rain effects mentioned earlier.

5.6.8 CORRELATION METHOD

In performing a correlation between SODAR wind speeds and mast (i.e., cup) wind speeds, a regression model is required. If we initially ignore the correction due to beam-drift, then the SODAR wind speed V_s can be expected to be a linear function of the cup wind speed V_c. Moreover, it is known that a SODAR will produce a wind speed estimate of zero when there is no wind, so a linear model without offset is most appropriate:

$$V_s = mV_c + \varepsilon, \tag{5.43}$$

where m is the slope and ε is the error, assumed to be random and normally distributed with zero mean and variance $\sigma_{V_s}^2$. Although there are errors associated with the cup measurements, orthogonal regression (allowing for errors in both V_c and V_s)

does not seem to offer new insights while inherently using the assumption that the cup errors are *equal* to the SODAR errors. Consequently, we use linear least-squares regression.

If N wind speeds V_{s_i}, V_{c_i} are available at a particular height, the residuals are

$$\varepsilon_i = V_{s_i} - m V_{c_i}, \quad i = 1, 2, \ldots, N$$

and least-squares minimizes $\displaystyle\sum_{i=1}^{N} \varepsilon_i^2$, to give an estimate

$$\hat{m} = \frac{\displaystyle\sum_{i=1}^{N} \frac{V_{c_i} V_{s_i}}{\sigma_i^2}}{\displaystyle\sum_{i=1}^{N} \frac{V_{c_i}^2}{\sigma_i^2}} \tag{5.44}$$

of the slope, with variance

$$\sigma_m^2 = \frac{1}{\displaystyle\sum_{i=1}^{N} (V_{c_i}^2 / \sigma_i^2)}. \tag{5.45}$$

Here σ_i^2 is the variance in V_{s_i}. There is evidence that availability, and hence $\sigma_{V_s}^2$, varies with height, but there is not strong evidence (e.g., from Fig. 5.36) for $\sigma_{V_s}^2$ depending on V_s. Hence we assume that the SODAR measurement error is independent of V_s at a particular height, and all $\sigma_i^2 = \sigma_{V_s}^2$.

The central calibration question is as follows. Given a SODAR wind speed measurement, V_s, what is the best estimate of the true wind speed, V, and what is the uncertainty, σ_V in that estimate?

From the above, the best estimate of the true wind speed is

$$\hat{V} = \frac{V_s}{\hat{m}}. \tag{5.46}$$

Also,

$$\sigma_V^2 = \left(\frac{\partial \hat{V}}{\partial V_s}\right)^2 \sigma_{V_s}^2 + \left(\frac{\partial \hat{V}}{\partial \hat{m}}\right)^2 \sigma_m^2$$

$$= \frac{\sigma_{V_s}^2}{\hat{m}^2} + \hat{V}^2 \frac{\sigma_m^2}{\hat{m}^2} \tag{5.47}$$

Note that

$$\sigma_{V_s}^2 \approx \frac{1}{N} \sum_{i=1}^{N} \varepsilon_i^2. \tag{5.48}$$

Finally, quality of regression is often judged by the correlation coefficient

$$r = \frac{\sum_{i=1}^{N}\left(V_{s_i} - \overline{V}_s\right)\left(V_{c_i} - \overline{V}_c\right)}{\sqrt{\sum_{i=1}^{N}\left(V_{s_i} - \overline{V}_s\right)^2 \sum_{i=1}^{N}\left(V_{c_i} - \overline{V}_c\right)^2}}$$

(5.49)

From (5.42), (5.44), (5.45), (5.48), and (5.49), it can be shown that

$$\frac{\sigma_m^2}{\hat{m}^2} = \frac{1-r^2}{r^2 N}\left(1 - \frac{\overline{V}_c^2}{(1/N)\sum_{i=1}^{N}V_{c_i}^2}\right) = \frac{1-r^2}{r^2 N}\frac{(1/N)\sum_{i=1}^{N}(V_{c_i} - \overline{V}_c)^2}{(1/N)\sum_{i=1}^{N}(V_{c_i} - \overline{V}_c)^2 + \overline{V}_c^2}.$$

If observed wind speeds are uniformly distributed between 0 and $V_{\max}$ then

$$\frac{\sigma_m^2}{\hat{m}^2} = \frac{1-r^2}{4r^2 N},$$

(5.50)

which is useful in relating correlation coefficient to uncertainty in slope. Similar equations are used to correlate wind directions.

5.6.9 DISTRIBUTION OF WIND SPEED DATA

During the calibration period the wind speeds will not be uniformly distributed (it could be a particularly windy period or a particularly calm period). Also the cup anemometers have a "starting wind speed" required to overcome their inertia and, cup data indicating wind speeds of less than 4 m s^{-1} are generally excluded because the cup calibration is considered unreliable. What are the implications for calibration of the probability distribution of wind speed?

For the linear model of (5.43), if $V_{s_i} \approx V_{c_i}$, as is expected here, then (5.44) predicts that the estimated slope m will not depend much on the distribution of wind speeds. On the other hand, (5.45) shows that σ_m^2 will be smaller if the wind speed distribution is more dominated by higher winds. Consequently, the inferences about the *quality* of the model fit (i.e., the uncertainty in the slope) would be expected to vary seasonally and depending on the duration of the calibration period. For the PIE case, the probability distribution is shown for 60 m height in Figure 5.42.

From (5.45) and (2.27),

$$\sigma_m^2 = \frac{\sigma_{V_s}^2}{N_{V_0}^2(1+2/q)}$$

for a Weibull distribution having scale parameter V_o and shape parameter k (Emeis, 2001). For example, from the PIE data above in Figure 5.42, if N were the same at both heights

$$\frac{\sigma^2_{m_{100}}}{\sigma^2_{m_{40}}}=\left(\frac{8.1}{9.7}\right)^2\frac{1+2/2.44}{1+2/2.34}=0.7$$

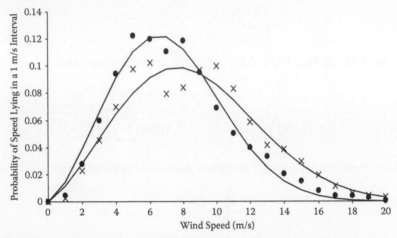

FIGURE 5.42 The wind speed probability at two heights during the calibration period. Solid circles: 40 m; crosses: 100 m. Solid lines: Weibull distribution fits with shape and scale parameters (2.44, 8.1 m s^{-1}) and (2.34, 9.7 m s^{-1}), respectively.

TABLE 5.3
Results of regression of SODAR wind speeds against mast wind speeds

System	Parameter	40 m	-60 m	80 m	100 m	120 m
AV4000	N	6580	6555	6281	5453	4676
	$\hat{m}$	1.082	1.085	1.083	1.079	1.080
	$\hat{m}/\hat{m}_{40}$	1	1.002	1.001	0.997	0.960
	σ_m	0.0009	0.0009	0.0009	0.0012	0.0016
	r^2	0.983	0.984	0.982	0.972	0.960
Metek	N	9454	9429	9408	8232	8292
	$\hat{m}$	0.944	0.935	0.928	0.923	0.936
	$\hat{m}/\hat{m}_{40}$	1	0.991	0.983	0.978	0.991
	σ_m	0.0012	0.0012	0.0012	0.0017	0.0014
	r^2	0.949	0.947	0.945	0.908	0.935
Scintec	N	6580	6555	6281	5453	4676
	$\hat{m}$	1.013	0.984	0.978	0.961	0.942
	$\hat{m}/\hat{m}_{40}$	1	0.971	0.966	0.949	0.930
	σ_m	0.0008	0.0008	0.0009	0.0010	0.0013
	r^2	0.982	0.982	0.979	0.977	0.965

giving not much difference in predicted slope errors. As seen from Table 5.3, there are around 20% fewer acceptable data points at 100 m compared to 40 m, so the diference in slope errors should be even smaller. Extreme differences in wind distribution, such as $(q, V_o) = (2, 5 \text{ m s}^{-1})$ and $(5, 15 \text{ m s}^{-1})$ would give a ratio of slope errors of around 0.2. For the PIE data, we would expect σ_m to be *less* at greater altitudes due to this effect, since wind speeds are generally higher.

5.6.10 REGRESSION SLOPE

Figure 5.43 shows correlations between each of the three SODARs and the mast cup anemometers at a number of heights corresponding to the sites on the mast shown in Figure 5.37. The regression results are summarized in Table 5.3. It may be seen that

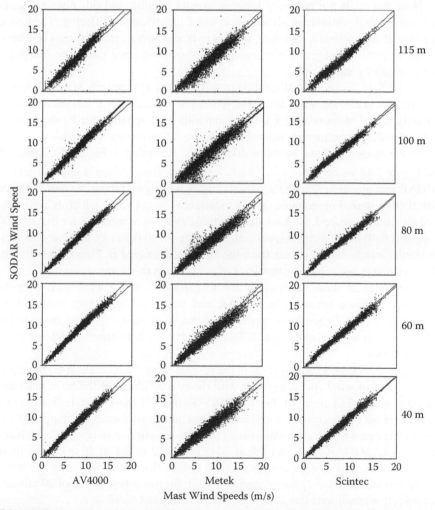

FIGURE 5.43 Regressions of SODAR wind speeds against mast wind speeds. Rows (from top): 120, 100, 80, 60, 40 m. Columns (from left): AV4000, Metek, Scintec.

1. scatter of data increases with height,
2. there are fewer data points and fewer high-wind points at greater heights,
3. slopes are not within 5%, and
4. correlation is high with values of ≥0.96.

The N values indicate that the Metek SODAR data may not have been filtered as strongly as the data from the other two SODARs. This is also suggested from the data spread in Figure 5.43. One consequence is that some fixed echo data have been included in the Metek regression at 100 m, as can be seen from low SODAR wind speeds compared to mast wind speeds. This also explains the significantly lower m value for that regression. For example, exclusion of points with $V_s < 3$ m s⁻¹ which also have $V_c-V_s > 2$ m s⁻¹ leads to a regression with $m = 0.9285$ instead of 0.923.

Note that σ_m is *not* generally lower at greater heights, and this does not agree with the Weibull estimation above, even when the effect of differing N values is included. It is concluded, therefore, that there is a genuine greater spread in data at greater heights. This is not really unexpected, however, since the lower data availability implies greater errors.

Figure 5.44 shows typical residual plots (ε versus V_c) for each SODAR. A running average is also shown in each case, where the average is over 100 points (monotonically sorted in increasing V_c). Variation with V_c is apparent for V_c above about 12 m s⁻¹ with the nonlinearity about 2% at 18 m s⁻¹. Referring to (4.14) this is entirely within the range predicted by beam drift effects. Referring to Eq. (5.45), Table 5.3, and Figure 5.44 we find that $\sigma_{V_s}^2 \gg \hat{V}^2\sigma_m^2$ and so $\sigma_V \approx \sigma_{V_s}$, which is not unexpected. SODAR manufacturers quote the uncertainty in wind speed measurement expected with their system. For example, Metek estimate $\sigma_{V_s} = 0.4\,\mathrm{m\,s^{-1}}$, which is consistent with the data in Figure 5.44. Also, the standard deviation of residuals for the AV4000 at 40 m is 0.40 m s⁻¹. Scintec specifications quote 0.1 to 0.3 m s⁻¹ accuracy for their horizontal winds, but it is clear from the data set displayed in Figure 5.44 that the SFAS system is not achieving that level of accuracy in this comparison.

How much of these residuals is due to variations in the wind itself over the separation distance between the SODAR and mast? In Figure 5.45, the residuals are plotted from a linear least-squares fit of the 80 m mast cup wind speed to the 60 m mast cup wind speed. Little difference is evident between this plot and those of Figure 5.44.

Figure 5.46 shows the rms error in the residuals for the AV4000 at 60 m height, as a function of wind speed, together with the rms residuals of the 80 m mast versus 60 m mast wind speed fit. For the SODAR–mast fit there is an indication of a small increase in uncertainty with increasing wind speed, but an estimate of 0.4 to 0.5 m s⁻¹ is again reasonable. This means that the variation in wind speed measured by mast and SODAR is around 4% at 10 m s⁻¹ and 2 to 3% at 20 m s⁻¹, for these 10-minute averages. Longer averages would reduce this error, providing the atmosphere was stationary over the averaging period: for times beyond about 20 minutes in convective conditions this assumption is probably not valid.

In Figure 5.46 the rms residuals for the mast–mast (i.e., cup–cup) comparisons are not distinctly different from the rms residuals for the SODAR–mast compari-

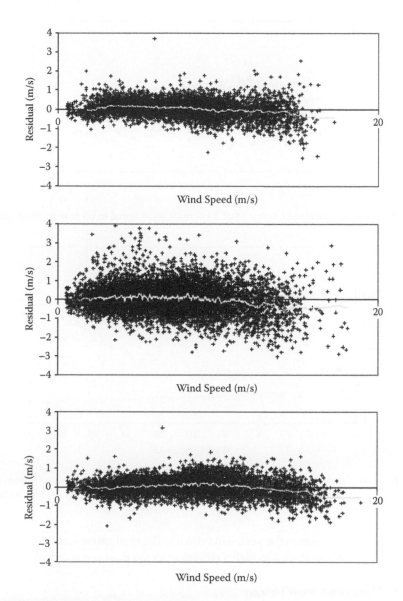

FIGURE 5.44 Residual plots for AV4000 (lower), Metek (center), and Scintec (upper) at 60 m. Superimposed line: running average of 100 points.

sons. For wind speeds up to 11 m s^{-1}, an F-test at the 95% level finds that the rms errors for the two fits are *not* significantly different.

This is an extremely important finding, since it suggests that there is *no difference* between SODAR–mast and mast–mast in terms of residuals, except at higher wind speeds. But importantly, the mast–mast comparison is between two sensors only 20 m apart, whereas the SODAR–mast comparison is for two sensors 70 m apart and not

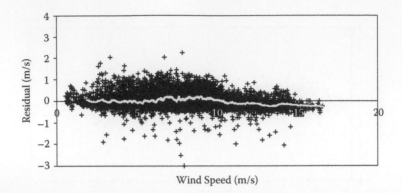

FIGURE 5.45   Residuals in a linear fit of 80 m mast wind speed to 60 m mast wind speed.

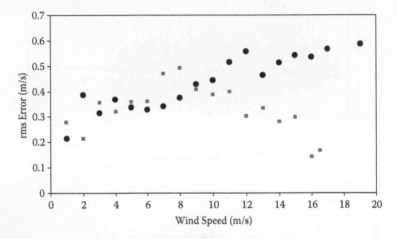

FIGURE 5.46   rms residual error (i.e., uncertainty in least-squares fitted wind speed) versus wind speed. Black circles: AV4000 versus cups at 60 m; small squares: cups at 80 m versus cups at 60 m.

necessarily exposed to even the same wind stream. The implication is that the SODAR is measuring winds to at least as high a reliability as the mast cup anemometers.

5.6.11 Variations with Height

The regression slopes, m, given in Table 5.3, should be independent of height if the SODAR is a well-designed wind-sensing tool and providing the calibrations have been conducted well. Figure 5.47 shows the slopes, or calibration coefficients, from Table 5.3. For the Metek, from (5.34) and (5.35) an out-of-level error would need to be at least 20° to explain the 6% calibration change. A leveling error of this magnitude would be easily visible just by inspecting the SODAR, and the level of the instrument was meticulously checked before and after the PIE field trials. For a leveling error with turning around the x-axis, Eq. (5.36) and the accompanying theory shows that

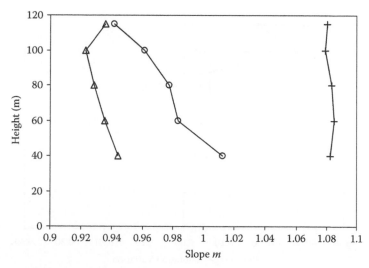

FIGURE 5.47   The calibration slope m as a function of height for the three SODAR systems. + = AV4000, Δ = Metek, O = Scintec.

$$V_s = \begin{pmatrix} 1 & 0 & 0 \\ 0 & \cos\alpha & -\sin\alpha \\ 0 & \sin\alpha & \cos\alpha \end{pmatrix} V_c$$

and so, because of the cos(α) dependence, out-of-level conditions *always* cause underestimation of the wind speed for a 3-beam system. This is a likely contribution to SODARs generally obtaining a calibration $m < 1$, although is unlikely to explain more than a few percent slope error.

This is because any increase in radial velocity on a tilted beam is more than cancelled by the increase in radial velocity on the normally vertical beam. This means that the calibration factor >1 for the AV4000 *cannot arise from out-of-level errors*. In the case of the AV4000, the calibration error of 8% is most likely due to incorrect estimation of the beam pointing angle by $\Delta\theta = 1.5°$.

If the SODAR measurements can be compared with wind speeds measured by well-calibrated cup anemometers at say 40 m, then it should be very easy to correct for any absolute calibration errors. This is shown in Figure 5.48 for the AV4000 system. With correction to 40 m, both the AV4000 and the Metek give wind speed estimates good to within about 2% at all heights.

The Scintec SODAR uses a combination of asymmetric opposing beams and a range of transmitted frequencies, with the lower frequencies being used preferentially for obtaining winds at greater heights. This could mean that the way in which data are handled is different at different heights (in the sense that different hardware is used and the software uses different parameters) and that this somehow causes the calibration change with height. A fixed echo from the mast could also possibly contaminate the combined spectral data. However, it must be emphasized that this result is from a particular field calibration and cannot be generalized to other situations.

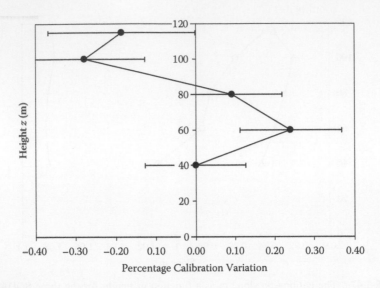

FIGURE 5.48 Expanded plot for the AV4000 showing the small variation of calibration with height.

5.6.12 Wind Direction Regressions

Monitoring wind direction is perhaps less important in wind energy applications, but regressions were also performed between SODAR directions and mast (wind vane) directions. Figures 5.49 and 5.50 show correlations between Metek wind directions and the vanes at 60 and 100 m on the Hovsoere mast. At 60 m the slope is 1.006 ± 0.0004 and $r^2 = 0.990$ (8528 points), and at 100 m the slope is 0.989 ± 0.003 and $r^2 = 0.891$ (3581 points). A fit through the origin for these data is a bit misleading, since the fit should be circular and repeat at 360°. Nevertheless, at 180° direction these fits predict errors of 1° at 60 m and 2° at 100 m, which are negligible for wind-energy support.

One curious artifact is that in the 100-m plot it is clear that for some points the sign of one of the individual wind components is wrong. This leads to a symmetric set of data points (e.g., mast −45°, SODAR +45°). It is not known what causes this occasional lapse, since it is not consistent with being due to fixed echoes.

5.7   SUMMARY

This chapter primarily discussed SODAR data quality and calibrations against standard wind sensors. Although many sources of error were investigated, nearly all can be ruled out as not significant.

The discussions in this chapter show how to minimize errors and conclude that one source of error could be significant: the absolute calibration of the beam pointing angle.

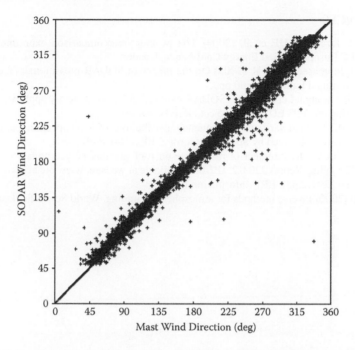

FIGURE 5.49 Regression of Metek-derived wind directions against the mast vane directions at 60 m.

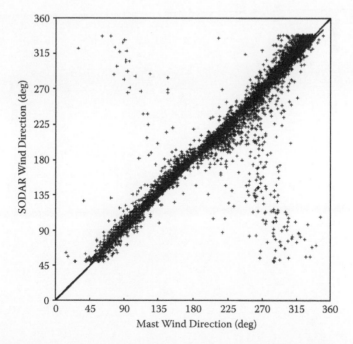

FIGURE 5.50 Regression of Metek-derived wind directions against the mast vane directions at 100 m.

REFERENCES

Antoniou I, Jørgensen HE et al. (2004) The profiler intercomparison experiment (PIE). EWEC European Wind Energy Conference, London.

Antoniou I, Jørgensen HE et al. (2005) On the theory of SODAR measurement techniques. Riso National Laboratory, 60 pp.

Bradley SG, Antoniou I et al. (2004) SODAR calibration for wind energy applications. Final reporting on WP3 EU WISE project NNE5-2001-297.

Emeis S (2001) Vertical variation of frequency distributions of wind speed in and above the surface layer observed by Sodar. Meteorol Z 10(2): 141–149.

Hopkins C (2004) Measurement of rain noise on roof glazing, polycarbonate roofing and ETFE roofing. Report 220312. IP2/06. Available at website: www.bre.co.uk/pdf/BRE–Report–220312.pdf, BRE Information Paper.

Rogers CD (2000) Inverse methods for atmospheric sounding. World Scientific, London.

6 SODAR Signal Analysis

In previous chapters we have described the atmospheric properties accessible to SODARs, elements of SODAR design, and instrument calibration. In a number of instances we have also discussed signal-to-noise ratio in general terms. In practice, separating valid signals from the noise background is a major part of SODAR hardware and software design. We consider these features in the current chapter.

6.1   SIGNAL ACQUISITION

6.1.1   SAMPLING

Although already discussed in Chapter 2, sampling will be briefly revisited. In the simplest case, a SODAR transmits a signal

$$A\sin(2\pi f_T t)$$

at a frequency f_T. The received signal is continuous, has reduced amplitude, in general is Doppler shifted and has modified phase

$$p(t) = A^* \sin\left(2\pi \left[f_T + \Delta f\right] t + \phi\right)$$

This signal can be *sampled* using an analog-to-digital converter (ADC) at times

$$t_m = m\Delta t = \frac{m}{f_s} \qquad m = 0, 1, \ldots$$

The *sampling frequency* is f_s. The sampled signal has discrete values

$$p_m = A^* \sin\left(2\pi m \frac{f_T + \Delta f}{f_s} + \phi\right) \tag{6.1}$$

6.1.2   ALIASING

For simplicity, write

$$\frac{f_T + \Delta f}{f_s} = n + \delta$$

where n is an integer 0, 1, …, and δ is a fraction. Then

$$p_m = A^* \sin\left(2\pi m \delta + \phi\right)$$

since $\sin(\theta \pm 2\pi mn) = \sin(\theta)$. As an example, assume $f_s = 960$ Hz: a signal component having frequency $960 + 960/3 = 1280$ Hz gives the same digitized values as if it had frequency $960/3 = 320$ Hz. The same is true for negative δ. This means that higher frequency components can add into the lower frequency spectrum. This is called *aliasing*. This means that all frequency components outside of $nf_s \pm f_s/2$ should be excluded from the signal before digitizing. This is called the *Nyquist criterion*. Usually this is interpreted as using *anti-aliasing* low-pass filters to remove all frequency components outside of $\pm f_s/2$, but in fact the criterion is satisfied if band-pass filters remove all components within a $\pm f_s/2$ *bandwidth* of nf_s.

6.1.3 MIXING

For a SODAR, the bandwidth of the Doppler spectrum is generally much smaller than f_s. For example, if the speed of sound is $c = 340$ m s^{-1}, $f_s = 4500$ Hz, and the beam tilt angle is $\pi/10$, the Doppler shift from a 10 m s^{-1} horizontal wind is $2 \times 10 \sin(\pi/10) \times 960/340 = 82$ Hz. So typically a filter need only have a bandwidth of, say, 200 Hz. It is usual to implement this filter as a low-pass filter, but this means that the signal frequencies of interest must lie below say 100 Hz, rather than be centered around, say, 4500 Hz. This is achieved by *demodulation* or *mixing down* the signal to be centered around 0 Hz.

The signal $p(t)$ is multiplied by a mixing waveform

$$M_I(t) = 2\sin(2\pi f_m t) \tag{6.2}$$

giving

$$M_I(t)p(t) = A^*\left\{\cos\left(2\pi\left[f_T - f_m + \Delta f\right]t + \phi\right) - \cos\left(2\pi\left[f_T + f_m + \Delta f\right]t + \phi\right)\right\}$$

This waveform has some frequency components centered around $f_T + f_m$ and others centered around $f_T - f_m$. If these groupings are well separated, then a low-pass filter can give just

$$I(t) = A^*\cos\left(2\pi\left[f_T - f_m + \Delta f\right]t + \varphi\right) \tag{6.3}$$

If the maximum negative value of Δf is less than $f_T - f_m$, then positive and negative Doppler shifts are then easily identified by looking at only the positive frequency part of the spectrum, as shown in Figure 6.1.

The frequency f_T sine wave generator is usually continuously running, but its output is switched to the speaker during the transmitted pulse. This is a very convenient signal to use as the mixing signal, so that $f_m = f_0$. However, since $\cos(2\pi\Delta f t) = \cos(2\pi[-\Delta f]t)$, positive and negative Doppler shifts cannot be distinguished. This means that, say, easterly and westerly winds will give the same result. To overcome this limitation, a *quadrature*, or 90° phase, signal is also mixed with the echo signal

$$M_Q(t) = 2\cos(2\pi f_T t)$$

giving

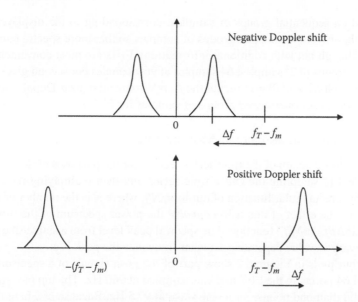

FIGURE 6.1 Positive and negative Doppler shifts are readily distinguished providing $f_T - f_m - |\Delta f| > 0$.

$$I(t) = A^* \cos(2\pi\Delta f\, t + \varphi)$$

$$Q(t) = A^* \sin(2\pi\Delta f\, t + \varphi)$$

(6.4)

This in-phase and quadrature-phase pair allows the amplitude, phase, and Doppler shift to be determined since

$$I(t) + jQ(t) = A^* e^{j(2\pi\Delta f\, t + \varphi)}$$

and the Fourier spectrum of this combination has either a single positive peak (for Δf positive, or a single negative peak (for Δf negative). Generation of a quadrature, cosine, signal at frequency f_T is generally a simple hardware task. The echo signal does need to be passed through two mixing circuits and sampled with two ADC channels.

The Doppler shift from a 20 m s^{-1} horizontal wind is 64 Hz for a 1.6-kHz SODAR, 180 Hz for a 4.5-kHz system, and 230 Hz for a 6-kHz system. It is necessary to sample at least twice the highest frequency, and depending on BP filter characteristics, perhaps three or four times the highest frequency. For example, the AeroVironment 4000 typically samples at 960 Hz, giving 960 s^{-1} × 400 m/340 m s^{-1} = 1130 samples for a height range of 200 m. In practice SODAR systems will usually sample a little longer than for the range displayed or recorded, to avoid combining echoes from more than one pulse. This also affords the opportunity to measure the background noise during the period at the top of the range in which no echoes are being returned. The total number of samples per pulse is not large, and so can be stored pending Fourier transforming. The fast Fourier transform (FFT) can be completed on sequential groups of samples, corresponding to the displayed range gate length, or from overlapped groups of samples so that more spectra can be displayed (although not with additional information). FFTs are most conveniently performed on groups of 2^n samples; 64 samples at 960 samples per second gives a range gate of $64 \times 340/(2 \times 960) = 11$ m, but the AeroVironment reports Doppler spectra at every 5 m (i.e., uses *overlapped* groups of samples for FFTs).

6.1.4   WINDOWING AND SIGNAL MODULATION

Sampling a finite length of the time series record, for the purposes of doing an FFT, is equivalent to sampling the entire time series and then multiplying the series of samples by a rectangular function of duration N/f_s where N is the number of samples in the FFT. The effect of this is to *convolve* the power spectrum of the time series with a $\sin(\pi N f/f_s)/(\pi N f/f_s)$ function. The spectral peak level from a single sine component will vary in value depending on what frequency the peak is at.

The four plots in Figure 6.2 show part of the positive half of a spectrum which contained 64 points in the FFT and was sampled at 960 Hz. The top plot shows the result (solid diamond points) for a sine wave at 97.5 Hz. Because of where the sampled points fall in relation to the peak of the $\sin(\pi N f/f_s)/(\pi N f/f_s)$ function, the resulting estimate of the peak is only 0.4 instead of 1.0. The second and third plots show results for sine waves at frequencies of 100 and 105.5 Hz. The bottom plot shows the result when the sampled time series has been multiplied by the Hanning window

$$H(t) = \frac{1}{2}\left[1 - \cos\left(2\pi \frac{f_s}{N-1} t\right)\right]$$

so that the sampled values always are small at the start of the sampled group and at the end. The result of this "windowing" is that the spectrum for a pure sine wave is wider (as shown in the ideal curve on the bottom plot). The worst-case position of the spectral peak with respect to the frequency bins then gives frequency estimates which are higher because they are on a wider curve. They are still only 0.7 instead of 1.0, however.

Other windows can be used: all give better estimation of peak value but poorer frequency resolution, when compared to the no-window case.

6.1.5   DYNAMIC RANGE

The amplified, filtered, and demodulated signal is an analog time series. This is fed to an ADC. The digital bit pattern is then stored as a representation of the sampled voltage of the SODAR signal. If the circuit has ramp gain to offset the spherical spreading loss, and has a band-pass filter to limit the noise bandwidth, then a 10-bit ADC is adequate. In this case, at best, the resolution is one part in 2^{10} (1:1024), or 0.1%. In practice, this is far more accurate than the generally noisy input signals. However, if no ramp gain is used, a SODAR signal could be expected to vary by at least a factor of $(320/10)^2 = 1024$ between heights 10 and 320 m. If 0.1% resolution is required at the upper height, then 20 bits are required. Thus to have a simpler

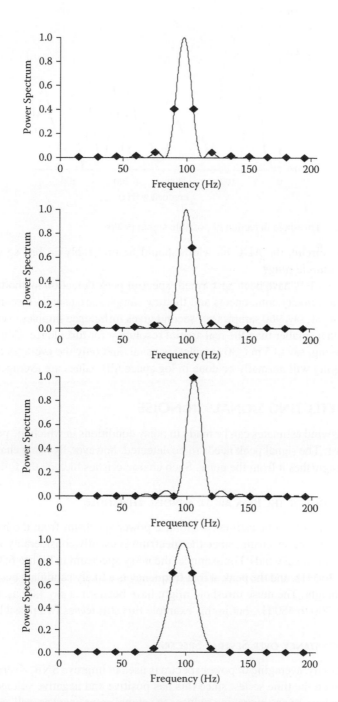

FIGURE 6.2 The effect of the Doppler shift not being a multiple of f_s/N.

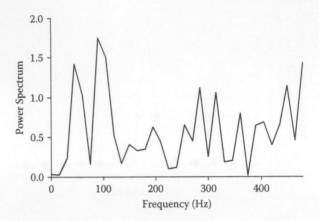

FIGURE 6.3 Threshold detection of possible signal peaks.

preamplifier circuit, the ADC bit width should be preferably 24 bits so as to have sufficient *dynamic range*.

Once the FFTs have been performed, spectral peak detection methods are used to determine velocity components and the raw samples are usually discarded. Note that sampling at, say, 960 samples per second gives turbulence samples every 0.18 m, which is much smaller than the real spatial resolution for turbulence. Consequently, some averaging, say to 5 m (~30 samples) is usual, and only the averages are stored. Such averaging will normally be done in log space (dB values are averaged).

6.2 DETECTING SIGNALS IN NOISE

Reasonable wind estimates can be made in noisy conditions in which the power SNR is less than 1. The signal peak needs to be detected, however, by some characteristic which distinguishes it from the noise. Such characteristics include the following.

6.2.1 HEIGHT OF THE PEAK ABOVE A NOISE THRESHOLD

Background noise can be estimated *within* a power spectrum from the highest frequency parts of the spectrum, since the spectrum is usually considerably wider than necessary for typical winds. For example, the noisy spectrum in Figure 6.3 has a signal peak at 100 Hz, and the peak at that frequency is a likely candidate because of its width and height. The noise threshold might have been set at say 1.0 based on noise levels from 300 to 480 Hz, but in this example this still leaves two possible peaks.

6.2.2 CONSTANCY OVER SEVERAL SPECTRA

Most commonly, averaging of power spectra is used to improve SNR. Averaging cannot be done on the time series, since this has positive and negative voltages and the phase is random, so any averaging reduces the signal component as well as the noise. But the power series is the square of the absolute value of the Fourier spectrum, and all phase information is therefore removed. Averaging the signal component does not

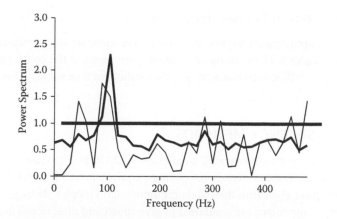

FIGURE 6.4　The effect of averaging the power spectrum shown in Figure 6.3.

change it, but averaging the noise component, which is random, reduces its fluctuations by the square root of the number of spectra in the average (see Figure 6.4).

For example, the AeroVironment 4000 typically records spectra at a particular range gate every 4 s, but displays data every five minutes. This means that 75 spectra are averaged. Taking the above example, and averaging successive spectra, gives the solid curve in Figure 6.4.

The peak position is often estimated from the *average* frequency in the spectrum (Neff, 1988):

$$\hat{f} = \frac{\int f P_R(f) df}{\int P_R(f) df}$$

(6.5)

but this should only be applied to the full, double-sided, spectrum.

6.2.3　NOT GENERALLY BEING AT ZERO FREQUENCY

In many circumstances it is known that there is some wind, and therefore any peak at zero frequency must be from a fixed echo. This part of the spectrum can then be ignored.

6.2.4　SHAPE

The spectrum shape for the signal component is often known from considerations of pulse length, etc. One way of discriminating against noise is to successively fit this shape with its peak at each spectral bin, and accept the position giving the best fit. A good approximation is a Gaussian, or even a parabola of the right width.

An even simpler variant is to take a weighted sum of several spectral bin values, and accept the position giving the highest sum. The weights can be all unity (searching for maximum power in a given signal BW), or reflect the expected shape of the signal peak.

6.2.5 SCALING WITH TRANSMIT FREQUENCY

A much more sophisticated method is to use two or more transmit frequencies. The Doppler shift scales with the transmit frequency, so peaks at the correct position in the spectra from different transmit frequencies indicate a true signal. This method is probably used by Scintec.

6.3 CONSISTENCY METHODS

Typically, the time series from a $\pm f_s/2$ bandwidth SODAR profile is sampled and FFTs performed on small blocks of samples, perhaps equivalent to 5 m vertically. A spectral peak detection algorithm then finds the individual Doppler shifts at each range gate. Velocity components are combined to give speed and direction. This results in individual and *independent* estimates of velocities at a series of vertical points.

Consistency checks and smoothing algorithms are then applied. This step makes a connection between the independent estimates (or assumes a connection). Combining velocity components may be interleaved with this check/smooth process.

Is it possible to come up with a *systematic* algorithm for smoothing, allowing for poor data points, and combining several profiles and points within a profile as consistency checks? The following method has been described by Bradley and Hünerbein (2004).

A typical plot of spectra versus height shows generally higher spectral peaks near the ground, and increasing spectral noise at higher altitudes. Examination of plots such as Figures 6.5 and 6.6 can indicate the most likely velocity profile by following the progression of spectral peaks with height.

At height z_m ($m = 1, 2, ..., M$), power spectral estimates $P_{im} = P(f_i, z_m)$ are measured at frequencies f_i ($i = 1, 2, ..., I$). The frequencies correspond to velocity compo-

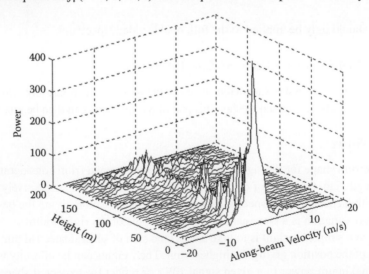

FIGURE 6.5 Typical raw power spectra versus height.

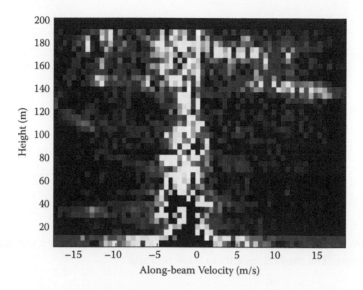

FIGURE 6.6 The spectra of Figure 6.5 shown as a contour plot.

nents u_i. Higher values of P_{im} are more likely associated with the echo signal rather than with noise. The quantity

$$\sigma^2_{im} = \frac{1}{P_{im}}$$

(6.6)

therefore represents the relative uncertainty of a particular f_i being at the signal peak for height z_m. We therefore treat the f_i, or equivalently the corresponding u_i, as measurements of signal peak position made with variance σ^2_{im}.

Assume that the u are a linear function of basis functions $K(z)$ with unknown coefficients x as follows.

$$u = Kx + \varepsilon$$

(6.7)

This puts the problem into the context of the solution of a set of linear equations. In particular, use of constraints, such as smoothness, profile rate of change, limiting the deviation from other data points, etc., can be applied by calling upon the huge constrained linear inversion literature.

There are still a number of **arbitrary decisions** required, however. These include

1. The relationship between the power spectral estimates and the variance,
2. The choice of basis functions, and
3. How to include other profile data as constraints.

Other possible relationships between P_{im} and σ^2_{im} include

The peak is the most likely estimator: $\sigma^2_{im} = \frac{1}{P^{\nu}_{im}}$

The center of a wider peak is a good estimator:

$$\sigma_{im}^2 = \frac{1}{\displaystyle\sum_{\nu=i-2}^{i+2} P_{\nu m}}$$

A fit to the peak gives the best estimator: $\sigma_{im}^2 = \displaystyle\sum_{\nu=1}^{I}\left[P_{\nu m} - P\left(u_i,u_\nu\right)\right]^2$

One example of basis function is a Gaussian

$$K_n\left(z\right) = \frac{1}{\sigma_z} e^{-\frac{1}{2\sigma_z^2}\left(z-\overline{z_n}\right)^2}$$

(6.8)

Figure 6.7 shows a typical fit using this method, but without any constraints from other profiles. The method appears to show promise.

6.4 TURBULENT INTENSITIES

There are two basic requirements in obtaining meaningful turbulent intensities:

1. Calibration of the system variable part of the SODAR equation and
2. Allowing for the background noise.

Calibration is actually quite difficult. One can try putting some well-defined scattering object above the SODAR, but this *must* be above the reverberation part

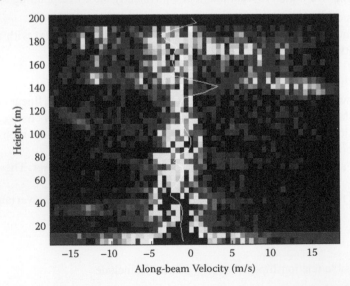

FIGURE 6.7   The fit through the spectra (white line) to give the spectral peak at each height. A Gaussian constraint is used for smoothness of velocity variations in the vertical.

of the SODAR range (i.e., above 20 m or so) and *must* be in the main beam of the SODAR (i.e., at 20 m the object must be located to within ±1 or 2 m horizontally). This is quite difficult with a tethered balloon, for example, but it might be possible to use an object on an overhead wire. Alternatively, a sonic anemometer can be used, providing one can work out how to extract meaningful records from it, and then allow for the extra vertical distance to the first usable SODAR range gate.

Background noise, P_N, can be allowed for using the turbulence or spectrum levels recorded from the highest one or two range gates, or from receiving without transmitting for a while, or from the wings of the power spectra. Then

$$C_T^2 = \frac{P - P_N}{\left[3 \times 10^{-4} \, P_T G A_e \tau \, c^{\frac{2}{3}} f_T^{\frac{1}{3}} \right] \left[\dfrac{e^{-2\alpha z}}{z^2 T^2} \right]}$$

(6.9)

If calibrated turbulence levels are required, care must also be exercised that fixed echoes are not contaminating the time series record. Gross fixed echoes are always evident on the SODAR facsimile display, but there is a problem with part contamination. So it is a good idea to look at the spectra on either side of a C_T^2 estimate, to see if there is a significant peak at zero frequency. The true signal spectral peak is of course also a measure of C_T^2, but this will be only available at the vertical spatial resolution of the winds, rather than the vertical spatial resolution of the turbulence: this reduced resolution may be adequate in many cases however.

C_V^2 measures derived from SODAR winds should be treated with caution: they will usually be only an approximation to the true values since assumptions are necessary on homogeneity and Taylor's "frozen field" hypothesis.

6.4.1 Second Moment Data

SODARs record σ_u, σ_v, and σ_w, the standard deviations of wind speed components. These standard deviations are useful as

1. An indicator of variability of winds (and likely uncertainties in u, v, and w),
2. Statistic variables to obtain other quantities such as wind energy, and
3. Input into similarity relationships to derive other quantities.

The latter is useful in, for example, obtaining estimates of **surface heat flux**, H, in convective conditions through (Weill et al. 1980)

$$\frac{\sigma_w^3}{z} = M \frac{H}{T}$$

(6.10)

where M is a constant and T is absolute temperature. Also, the **mixing layer height**, Z_m, can be estimated through (Asimakopoulos et al., 2002)

$$\frac{d\sigma_w^2}{dz} = 0 \text{ at } z = \frac{Z_m}{3.2}$$

(6.11)

6.5   PEAK DETECTION METHODS OF AEROVIRONMENT AND METEK

The SODAR incorporates signal-processing software to determine

1. The position in the spectrum of the signal peak (corresponding to Doppler shift) and
2. The averages over a number of profiles (to improve SNR).

   The methods for achieving these tasks vary a little between manufacturers. Some examples follow.

6.5.1   AeroVironment

The AeroVironment system performs peak detection on each individual 64-point spectrum (128-point spectra can also be user-selected). This is done by finding the highest power in any contiguous 5-spectral-point group (or 7-point for a 128-point spectrum) across the frequency spectrum. The SNR is then defined as the 5-point power divided by the power in the remaining 59 points normalized by multiplying by 59/5. Finally, averaging the accepted peak positions over N_s profiles gives the estimated Doppler shift for the particular range gate and beam. Note that if the user selects the option to use beam 3 data, then a rejected beam 3 spectrum causes the beam 1 and beam 2 peak estimates to also be rejected at that range gate for that profile (i.e., the system does not default to a 2-beam configuration which might give averages of mixed 2-beam and 3-beam calculations). Numbers of accepted beam 1, beam 2, and beam 3 peak estimates in each averaging interval are output for the user.

   The system also employs an adaptive noise threshold as part of the decision to accept/reject a spectrum. This threshold is determined by sampling the background noise prior to the transmit pulse, and appropriately scaling this threshold to account for spherical beam divergence with altitude. This option can be disabled or enabled by the user. If this option is disabled, the system uses a fixed noise threshold which is applied at every altitude.

   Statistical analysis shows that the uncertainty in each estimate of the position of the spectral peak in this scheme depends on

$$(SNR_1)\left(\frac{\Delta f}{\sigma_f}\right) .$$

6.5.2   Metek

Metek average N_s spectra for each beam and each range gate. Each recorded value in a spectrum is the sum $(P_A + P_N)$ of the echo P_A from atmospheric turbulence and the Gaussian noise P_N which has zero mean and variance. The SNR from a single spectral estimate is

$$SNR_1 = \frac{P_A}{\sigma_P}$$

If N_s spectra are averaged, the average spectral estimate becomes $P_A + \dfrac{1}{N_s}\displaystyle\sum_1^{N_s} P_N$,

and the variance in this estimate is $\mathrm{var}\left(\dfrac{1}{N_s}\displaystyle\sum_1^{N_s} P_N\right) = \dfrac{1}{N_s}\sigma_P^2$. The SNR is therefore

$$SNR_{N_s} = \sqrt{N_s}\,\frac{P_A}{\sigma_P}$$

(6.12)

In the Metek SODAR, 32 complex Fourier amplitudes are obtained over $N_s = 20$ to 60 profiles, giving 32 averaged spectral intensities. Two noise spectra measurements are made shortly before each pulse is transmitted and these are averaged to obtain an estimate of P_N at each frequency in the Fourier spectrum. These averaged noise intensities are subtracted from the averaged intensities received after the pulse, to give residual power spectra at each range gate. It is assumed that the noise-free signal power spectrum has a Gaussian shape

$$\frac{P_0 \Delta f}{\sqrt{2\pi}\sigma_f}\, e^{-\frac{1}{2}\left(\frac{f-\hat{f}}{\sigma_f}\right)^2}$$

where $\Delta f = \dfrac{1}{T}$ is the frequency resolution. If logarithms of the spectral estimates are used,

$$\ln\left(\frac{P_0 \Delta f}{\sqrt{2\pi}\sigma_f}\right) - \frac{1}{2}\left(\frac{f-\hat{f}}{\sigma_f}\right)^2$$

(6.13)

is a quadratic in f. Using least-squares, the moments P_0, $\hat{f}$, and σ_f can be estimated. In practice, only n spectral points within 1/4 height (6 dB) of the main peak are included in the least-squares fit. Simulations based on this scheme show that, for high SNR and with $N_s > 40$, the uncertainty in the peak position $\hat{f}$ is about 0.06 spectral bin widths and the uncertainty in σ_f is about 0.2 spectral bins. If all cases are rejected which have SNR below a certain critical threshold, then this accuracy is expected. With $\Delta z = 20$ m and $f_T = 1675$ Hz, the error in the radial velocity component is $\sigma_{v_r} = \sigma_f \Delta z = 0.1 \mathrm{m\ s}^{-1}$ and the error in the estimate of the width of the velocity spectrum is 0.17 m s⁻¹. For a tilt angle of $\theta = 20°$, and given that the two horizontal velocity components are generally comparable and dominate over the vertical component,

$$\sigma_V \approx \frac{\sqrt{2}\sigma_{v_r}}{\sin\varphi} = 0.4\mathrm{m\ s}^{-1}.$$

Similar analysis gives the uncertainty σ_ψ in the wind direction as about 6° for $V = 5$ m s⁻¹.

6.6   ROBUST ESTIMATION OF DOPPLER SHIFT FROM SODAR SPECTRA

6.6.1   FITTING TO THE SPECTRAL PEAK

Assume that a sinusoidal signal $s(t)$ of duration τ is transmitted. The *amplitude spectrum* of the received voltage is

$$V_i \approx \overline{V}_i + E_i \tag{6.14}$$

where $\overline{V}_i$ is the received scattered signal component and E_i arises from random noise. For Gaussian-distributed E_i, the probability of recording a spectral amplitude magnitude between and $|V_i|$ and $|V_i| + d|V_i|$ is

$$p\left(|V|_i\right) d|V_i| = \frac{1}{\sqrt{2\pi}\sigma_E} e^{-\frac{1}{2}\left(\frac{|V_i| - |\overline{V}_i|}{\sigma_E}\right)^2} dV_i \tag{6.15}$$

where σ_E^2 is the variance of E_i. The power spectral estimate at f_i is $P_i = V_i^* V_i = |V_i|^2$ so

$$p(P_i) = p\left(|V_i|\right)\frac{|dV_i|}{dP_i} = \frac{1}{2\sqrt{2\pi P_i}\sigma_E} e^{-\frac{1}{2}\left(\frac{\sqrt{P_i} - |\overline{V}_i|}{\sigma_E}\right)^2} \tag{6.16}$$

From this probability distribution, the mean power spectral value at frequency f_i is

$$\overline{P}_i = \int_0^\infty P_i p(P_i) dP_i = \frac{1}{\sqrt{2\pi}\sigma_E} \int_{-\infty}^\infty |V_i|^2 e^{-\frac{1}{2}\left(\frac{|V_i| - |\overline{V}_i|}{\sigma_E}\right)^2} d|V_i| = |\overline{V}_i|^2 + \sigma_E^2 \tag{6.17}$$

In other words, there is a systematic *overestimate* of the power spectral value by the noise power quantity

$$\overline{N} = \sigma_E^2 \tag{6.18}$$

Consequently, we subtract from the spectrum an estimate, $\hat{N}$, of the mean power level when no signal is present (i.e., from the highest range gates) giving a *reduced power spectrum*

$$P_i' = P_i - \hat{N} \tag{6.19}$$

The moments of are

$$\hat{\overline{N}} = \overline{N}$$

$$\sigma_{\hat{N}}^2 = \frac{3\overline{N}^2}{N_f N_{av}}$$

(6.20)

where N_{av} are the number of spectra which are averaged to obtain $\hat{N}$.
This results in moments

$$\overline{P'_i} = \overline{P_i} - \hat{\overline{N}} = \overline{|V_i|}^2$$

$$\sigma_{P'_i}^2 = \sigma_{P_i}^2 + \sigma_{\hat{N}}^2 = 2\overline{N}\left(2\overline{|V_i|}^2 + \overline{N}\right) + \frac{3\overline{N}^2}{N_f N_{av}} \approx 2\overline{N}\left(2\overline{P'_i} + \overline{N}\right)$$

(6.21)

Various pulse envelope shapes are used, but all allow $\overline{P'_i}$ to be represented in the form

$$\overline{P'_i} \approx P_{max} e^{-\frac{1}{2\sigma_f^2}(f_i - f_D)^2}$$

(6.22)

Then

$$\ln\frac{P_{ref}}{\overline{P'_i}} = \left(\frac{f_D^2}{2\sigma_f^2} - \ln\frac{P_{max}}{P_{ref}}\right) + \left(-\frac{f_D}{\sigma_f^2}\right)f_i + \left(\frac{1}{2\sigma_f^2}\right)f_i^2$$

(6.23)

where P_{ref} is a reference value (for example, 1 V^2). In other words, the logarithm of the reduced power spectrum has a quadratic dependence on frequency.

We find the nearest spectral frequency to the peak position, and write the index i relative to this, so the nearest spectral frequency to the peak is labeled f_0. Least-squares is used to estimate the three coefficients of the quadratic using $2Q+1$ points centered around f_0 (typically $Q = 2$ or 3). We use an odd number of fitting points because in the case of unweighted least-squares this leads to simplification.

We now apply the above methods to raw spectral data recorded from a Metek SODAR/RASS. The relevant parameters are given in Table 6.1. The time-series echo strength is recorded for 3.2 s and each range gate (region over which each spectrum is valid) is 16 m in vertical extent. The atmospheric conditions were low wind and fairly neutral conditions (so relatively weak reflections) but with low levels of external background acoustic noise.

Figure 6.8 shows a typical Hanning-windowed time series for one range gate. Figure 6.9 shows the corresponding amplitude spectrum. Note the signal peak near the transmitting frequency. From such spectra more localized spectra are selected, so that only possible Doppler shifts are included in the analysis. For a beam tilt angle of 20°, the radial velocity component for a horizontal wind of 20 m s^{-1} will give a

TABLE 6.1

List of parameters for the Metek SODAR

Parameter	Description	Value
f_T (Hz)	Transmitted frequency	1674
τ (s)	Pulse duration	0.0958
f_s (Hz)	Sampling frequency	44100
N_f	Number of spectral estimates	4096
Δf (Hz)	Frequency interval in spectrum	10.8

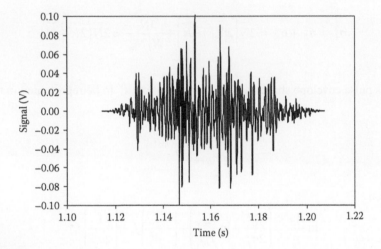

FIGURE 6.8 A typical Hanning-windowed time series for one range gate from the Metek SODAR.

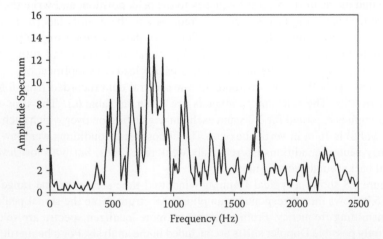

FIGURE 6.9 The spectrum for a single range gate Hanning-windowed time series.

Doppler shift of 67 Hz, so considering 16 spectral frequencies over the range 1593 to 1755 Hz should suffice for this data set. Figure 6.10 shows a power spectrum (P_i values) from the range gate centered at 197 m. Estimation of SNR using the wings of the spectrum around the peak value gives SNR = 8 dB. In Figure 6.11 the data values for the quadratic fit are shown. The estimated value for peak frequency is f_D =1681±2 Hz and for width σ_f = 9.7±0.9 Hz.

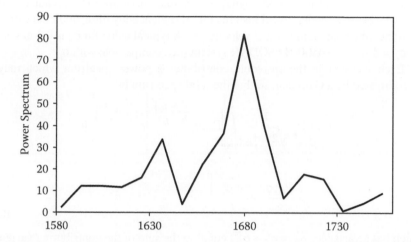

FIGURE 6.10 A local spectrum taken from range gate 13 (height 197 m) and for which the estimated SNR is 8 dB.

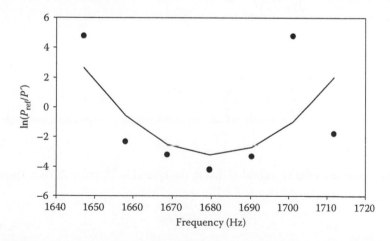

FIGURE 6.11 Plot of log-corrected power spectrum from data in Figure 6.10 and with $Q = 3$. Data are shown with dark dots and the fit with a solid line.

6.6.2 ESTIMATION OF σ_w

In practice reflections are from an ensemble of scatterers which provide a continuum of Doppler shifts. This gives spread to the Doppler spectrum which is particularly important for vertical profiling since the variance, σ_w^2, in vertical velocity is an important boundary layer parameter.

Assume that the Doppler frequency from the ensemble has a Gaussian probability centered on $\overline{f}_D$ and with standard deviation σ_D. This range of Doppler frequencies will cause spectral broadening of the signal, and estimation of this extra broadening from a *vertical* beam provides useful insights into turbulent eddy dissipation rates through the standard deviation in vertical velocity, σ_w. A typical value for σ_w is 0.3 m s^{-1}, giving $\sigma_D = 8$ Hz for a 4500-Hz SODAR system (i.e., comparable with σ_f).

Each scatterer in the ensemble contributes a power spectrum which may be approximated by a Gaussian, so that the total spectrum is

$$\overline{P'}(f) = \frac{1}{\sqrt{2\pi}\,\sigma_D} \int_{-\infty}^{\infty} \left[P_{\max} e^{-\frac{1}{2}\left(\frac{f-f_D}{\sigma_f}\right)^2} \right] e^{-\frac{1}{2}\left(\frac{f_D-\overline{f}_D}{\sigma_D}\right)^2} \, df_D$$

$$= \frac{P_{\max}\sigma_f}{\sigma_T} e^{-\frac{1}{2}\left(\frac{f-\overline{f}_D}{\sigma_T}\right)^2} \tag{6.24}$$

which has a variance, $\sigma_T^2 = \sigma_f^2 + \sigma_D^2$, equal to the sum of the contributing variances, as expected. There is now an extra variability (in addition to the background noise discussed above) given by

$$\sigma_{P'}^2 = \frac{1}{\sqrt{2\pi}\,\sigma_D} \int_{-\infty}^{\infty} \left[P_{\max} e^{-\frac{1}{2}\left(\frac{f-f_D}{\sigma_f}\right)^2} \right]^2 e^{-\frac{1}{2}\left(\frac{f_D-\overline{f}_D}{\sigma_D}\right)^2} \, df_D - \overline{P'}^2$$

or

$$\frac{\sigma_{P'}^2}{\overline{P'}^2} = \frac{1 + \dfrac{\sigma_D^2}{\sigma_f^2}}{\sqrt{1 + \dfrac{2\sigma_D^2}{\sigma_f^2}}} e^{\frac{\sigma_D^2}{\left(\sigma_f^2 + 2\sigma_D^2\right)\sigma_T^2}\left(f - \overline{f}_D\right)^2} - 1 \tag{6.25}$$

The maximum relative variation due to the spread in Doppler shifts is therefore

$$\left. \frac{\sigma_{P'}^2}{\overline{P'}^2} \right|_{\max} = \frac{1+\alpha}{\sqrt{1+2\alpha}} - 1 \tag{6.26}$$

where $\alpha = \dfrac{\sigma_D^2}{\sigma_f^2}$.

The regression methods discussed above can be used to estimate $\sigma_T^2 = \sigma_f^2 + \sigma_D^2$ and hence σ_D since σ_f^2 is known from the system design. The relative error in estimated σ_D is

$$\frac{\sigma_{\sigma_D}}{\sigma_D} = \frac{\sigma_{\sigma_T}}{\sqrt{\sigma_T^2 - \sigma_f^2}} = \frac{1}{\sqrt{1 - \dfrac{\sigma_f^2}{\sigma_T^2}}} \frac{\sigma_{\sigma_T}}{\sigma_T}$$

(6.27)

If $\sigma_D^2 << \sigma_f^2$ there is a large relative error multiplication factor in (6.27). Figure 6.12 shows the relative error in σ_w as a function of σ_w for several values of SNR and $\dfrac{f_T}{\sigma_f}$.

This emphasizes the importance of having good SNR for σ_w estimates, as well as the preference of a long pulse (small σ_f). Higher frequency SODARs also do better. A value of

$$\frac{f_T}{\sigma_f} = 170$$

corresponds approximately to a 1700 Hz SODAR having a 0.05-s pulse. For large σ_w, the relative **error** asymptotically approaches $\dfrac{\sigma_{\sigma_T}}{\sigma_T}$.

6.7   AVERAGING TO IMPROVE SNR

The **time series** from successive profiles should not be averaged, since they are incoherent and will average toward zero.

Averaging of **power spectra** from successive profiles is useful, since phase information has been removed. The noise power fluctuates more than the signal,

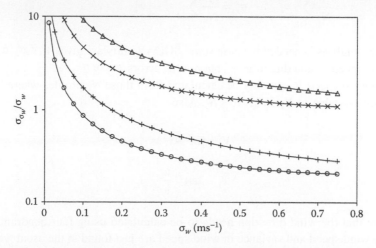

FIGURE 6.12   Relative error in sigma-w value. Five-point fits with peak at a spectrum frequency. SNR = 20 dB, f_T/σ_f = 170 (plus signs); SNR = 20 dB, f_T/σ_f = 340 (circles); SNR = 10 dB, f_T/σ_f = 170 (triangles); SNR = 10 dB, f_T/σ_f = 340 (crosses).

providing the averaging time is not too long (say no longer than 20 minutes, but this signal autocorrelation time will depend on the environment). Noise powers P_{N_i} from the ith profile, at a particular range gate, are summed in the averaging process

$$\overline{P_N} = \frac{1}{n}\sum_{i=1}^{n} P_{N_i}$$

(6.28)

and

$$\sigma_{av}^2 = \sum_{i=1}^{n}\left(\frac{\partial \overline{P_N}}{\partial P_{N_i}}\right)^2 \sigma_{P_{N_i}}^2 = \sigma_{P_N}^2 \sum_{i=1}^{n}\left(\frac{1}{n}\right)^2 = \frac{\sigma_{P_N}^2}{n}$$

(6.29)

so the standard deviation of the noise goes down as the square root of the number of averages.

6.7.1 Variance in Wind Speed and Direction over One Averaging Period

Generally wind data from a number of profiles are averaged. In the following we will restrict attention to the horizontal wind components. The ith profile may contain an acceptable u_i wind component and/or an acceptable v_i component. This results, after an averaging period, in N_u east-west components and N_v north-south components. The means and variances from a *single* averaging period are

$$\overline{u} = \frac{1}{N_u}\sum_{i=1}^{N_u} u_i \qquad \sigma_u^2 = \frac{1}{N_u}\sum_{i=1}^{N_u}\left(u_i - \overline{u}\right)^2 = \frac{1}{N_u}\sum_{i=1}^{N_u} u_i^2 - \left(\overline{u}\right)^2$$

$$\overline{v} = \frac{1}{N_v}\sum_{i=1}^{N_v} v_i \qquad \sigma_v^2 = \frac{1}{N_v}\sum_{i=1}^{N_v} v_i^2 - \left(\overline{v}\right)^2$$

(6.30)

Some analysis is needed because some SODAR software gives $\overline{u}, \overline{v}, \sigma_u, \sigma_v$, and the mean speed $\overline{V}$ and direction $\overline{\psi}$, but not the errors $\sigma_{\overline{v}}$ or $\sigma_{\overline{\psi}}$.

The wind speed V_i can only be calculated from those N_V profiles where both u_i and v_i are available so $N_V \leq N_u$, $N_V \leq N_v$. Also

$$V_i = \left(u_i^2 + v_i^2\right)^{\frac{1}{2}}$$

$$\Psi_i = \tan^{-1}\frac{u_i}{v_i}$$

(6.31)

Note that the wind direction needs to be calculated using four quadrants. The average wind speed and variance in wind speed are just found in the usual way

$$\overline{V} = \frac{1}{N_V}\sum_{i=1}^{N_V}\left(u_i^2 + v_i^2\right)^{\frac{1}{2}} \qquad \sigma_V^2 = \frac{1}{N_V}\sum_{i=1}^{N_V}\left(u_i^2 + v_i^2\right) - \left(\overline{V}\right)^2$$

(6.32)

or

$$\sigma_V^2 \approx \frac{N_u}{N_V}\left[\sigma_u^2 + \left(\overline{u}\right)^2\right] + \frac{N_v}{N_V}\left[\sigma_v^2 + \left(\overline{v}\right)^2\right] - \left(\overline{V}\right)^2$$

(6.33)

Also,

$$\sigma_{\overline{V}}^2 \approx \frac{\sigma_V^2}{N_V} = \frac{N_u}{N_V^2}\left[\sigma_u^2 + \left(\overline{u}\right)^2\right] + \frac{N_v}{N_V^2}\left[\sigma_v^2 + \left(\overline{v}\right)^2\right] - \frac{1}{N_V}\left(\overline{V}\right)^2$$

(6.34)

is the variance in the mean wind speed over the averaging period.

The direction needs to be found from the accumulated wind runs in each component, since otherwise averaging could result in a nearly 0° direction being interpreted as nearly 180°. So

$$\overline{\psi} \approx \tan^{-1}\frac{\overline{u}}{\overline{v}}$$

(6.35)

This is why, for the AeroVironment SODAR, no "number of recorded values" is given for the direction.

The variance in direction is

$$\sigma_{\overline{\psi}}^2 = \sum_{i=1}^{N_u}\left(\frac{\partial\overline{\psi}}{\partial u_i}\right)^2\sigma_u^2 + \sum_{i=1}^{N_v}\left(\frac{\partial\overline{\psi}}{\partial v_i}\right)^2\sigma_v^2$$

$$= \sum_{i=1}^{N_u}\left(\frac{\partial\tan\overline{\psi}}{\partial u_i}\frac{1}{1+\tan^2\overline{\psi}}\right)^2\sigma_u^2 + \sum_{i=1}^{N_v}\left(\frac{\partial\tan\overline{\psi}}{\partial v_i}\frac{1}{1+\tan^2\overline{\psi}}\right)^2\text{(}$$

$$= \left(\frac{\left(\overline{v}\right)^2}{\left(\overline{u}\right)^2 + \left(\overline{v}\right)^2}\right)^2\frac{1}{\left(N_v\overline{v}\right)^2}\left[N_u\sigma_u^2 + N_v\sigma_v^2\left(\frac{N_u\overline{u}}{N_v\overline{v}}\right)^2\right]$$

$$= \frac{N_u}{N_v}\frac{\sigma_u^2 + \frac{N_u}{N_v}\sigma_v^2\tan^2\overline{\psi}}{N_v\left(\overline{v}\right)^2\left[\tan^2\overline{\psi}+1\right]^2}$$

(6.36)

6.7.2 Combining Wind Data from a Number of Averaging Periods

For wind speed S, and wind direction ψ,

$$S = \sqrt{u^2 + v^2}$$

(6.37)

$$\psi = \tan^{-1}\frac{u}{v}$$

(6.38)

where u and v are the vector components.

We assume there are measurements u_i, v_i, $i=1,2, \ldots, N$ from N profiles, where the u_i and v_i are measured with individual uncertainties σ_{u_i} and σ_{v_i}. Assume that these uncertainties arise from taking the mean of n_{u_i} values of u, and n_{v_i} values of v, each with variance σ_1^2, so that

$$\sigma_{u_i}^2 = \frac{\sigma_1^2}{n_{u_i}}$$

(6.39)

$$\sigma_{v_i}^2 = \frac{\sigma_1^2}{n_{v_i}}$$

(6.40)

where σ_1^2 arises from error in estimating the position of the spectral peak at each range gate, and is essentially the same for each estimation.

Now

$$\sigma_{S_i}^2 = \left(\frac{\partial S_i}{\partial u_i}\right)^2 \sigma_{u_i}^2 + \left(\frac{\partial S_i}{\partial v_i}\right)^2 \sigma_{v_i}^2$$

$$= \left[\frac{1}{n_{u_i}}\left(\frac{u_i}{S_i}\right)^2 + \frac{1}{n_{v_i}}\left(\frac{v_i}{S_i}\right)^2\right]\sigma_1^2$$

$$= \frac{\sigma_1^2}{\alpha_i}$$

(6.41)

is the variance of a single S_i, and

$$\sigma_{\psi_i}^2 = \left(\frac{\partial \psi_i}{\partial u_i}\right)^2 \sigma_{u_i}^2 + \left(\frac{\partial \psi_i}{\partial v_i}\right)^2 \sigma_{v_i}^2$$

$$= \left[\frac{1}{n_{u_i}}\left(\frac{v_i}{S_i^2}\right)^2 + \frac{1}{n_{v_i}}\left(\frac{u_i}{S_i^2}\right)^2\right]\sigma_1^2$$

$$= \frac{\sigma_1^2}{\beta_i}$$

(6.42)

is the variance of a single ψ_i.

The mean $\overline{S}$ and $\overline{\psi}$ are required over the N measurements, allowing for the variable uncertainties. These means are found by following the usual procedures for modeling $y = a + bx$, but here we have only one parameter $a = \overline{y}$, so the one-parameter weighted least-squares fit has the form $y = \overline{y}$.

The single parameter, $\overline{y}$, is found by minimizing

$$\chi^2 = \sum_i \left(\frac{y_i - \bar{y}}{\sigma_i} \right)^2$$

(6.43)

where σ_i^2 is the variance in measurement y_i, giving

$$\bar{y} = \frac{1}{\sum_i \frac{1}{\sigma_i^2}} \sum_i \frac{y_i}{\sigma_i^2}$$

(6.44)

and

$$\sigma_{\bar{y}}^2 = \frac{N}{\sum_{i=1}^{N} \frac{1}{\sigma_i^2}}$$

(6.45)

In the context of wind-averaging of $N=10$ one-minute values, this gives

$$\bar{S} = \frac{1}{\sum_{i=1}^{10} \alpha_i} \sum_{i=1}^{10} \alpha_i S_i$$

(6.46)

and

$$\bar{\psi} = \frac{1}{\sum_{i=1}^{10} \beta_i} \sum_{i=1}^{10} \beta_i \psi_i$$

(6.47)

where the weights are

$$\alpha_i = \left[\frac{1}{n_{u_i}} \left(\frac{u_i}{S_i} \right)^2 + \frac{1}{n_{v_i}} \left(\frac{v_i}{S_i} \right)^2 \right]^{-1}$$

(6.48)

and

$$\beta_i = \left[\frac{1}{n_{u_i}} \left(\frac{v_i}{S_i^2} \right)^2 + \frac{1}{n_{v_i}} \left(\frac{u_i}{S_i^2} \right)^2 \right]^{-1}$$

(6.49)

Similar considerations can be used for any other averaged quantities.

An example taken from an AeroVironment 4000 return from 90 m with averaging over 150 s, has measured values of $u_i = -3.4$ m s^{-1}, $\sigma_{u_i} = 0.8$ m s^{-1}, $n_{u_i} = 38$, $v_i = 3.7$ m s^{-1}, $\sigma_{v_i} = 0.9$ m s^{-1}, and $n_{v_i} = 36$. This gives $S_i = 5.0$ m s^{-1}, $\psi_i = 313°$, and $\sigma_1 = 5$ m s^{-1}. Then $\alpha_i = 36$ and $\beta_i = 920$ rad^{-2}m^{2}s^{-2}. This means that the standard deviation in wind speed for this averaging period is $\sigma_{S_i} = 0.83$ m s^{-1} and the standard deviation in wind direction is $\sigma_{\psi_i} = 9.5°$.

6.7.3 DIFFERENT AVERAGING SCHEMES FOR SODAR AND STANDARD CUP ANEMOMETERS

Cup anemometers represent one "standard" against which SODARs might be calibrated. As pointed out by Antoniou and Jørgensen (2003) cup anemometers measure wind run and divide by averaging time to obtain wind speed. Thus

$$V_{cup} = \frac{1}{T} \int_0^T V dt = \frac{1}{T} \int_0^T \sqrt{u^2 + v^2}\, dt$$

(6.50)

whereas a SODAR obtains wind speed from the averaged u and the averaged v components:

$$V_{SODAR} = \sqrt{\left(\frac{1}{T} \int_0^T u dt\right)^2 + \left(\frac{1}{T} \int_0^T v dt\right)^2}$$

(6.51)

To allow for the sampled nature of the SODAR (a sample each profile), assume that the wind is essentially in the +x-direction with small perturbations:

$$u = U + u_i$$

$$(i-1)\Delta t < t \le i\Delta t$$

$$v = v_i$$

(6.52)

Then

$$V_{cup} = \frac{1}{N} \sum_{i=1}^{N} \sqrt{\left(U+u_i\right)^2 + v_i^2}$$

$$= \frac{U}{N} \sum_{i=1}^{N} \sqrt{1 + \frac{2u_i}{U} + \frac{u_i^2 + v_i^2}{U^2}}$$

$$\approx \frac{U}{N} \sum_{i=1}^{N} \left(1 + \frac{u_i}{U} + \frac{u_i^2 + v_i^2}{2U^2}\right)$$

$$\approx U + \frac{1}{N} \sum_{i=1}^{N} u_i + \frac{1}{2NU} \sum_{i=1}^{N} \left(u_i^2 + v_i^2\right)$$

(6.53)

and

$$V_{SODAR} = \sqrt{\left(U + \frac{1}{N} \sum_{i=1}^{N} u_i\right)^2 + \left(\frac{1}{N} \sum_{i=1}^{N} v_i\right)^2}$$

$$= U \sqrt{1 + \frac{2}{N} \sum_{i=1}^{N} \frac{u_i}{U} + \left(\frac{1}{N} \sum_{i=1}^{N} \frac{u_i}{U}\right)^2 + \left(\frac{1}{N} \sum_{i=1}^{N} \frac{v_i}{U}\right)^2}$$

$$\approx U \left[1 + \frac{1}{N} \sum_{i=1}^{N} \frac{u_i}{U} + \frac{1}{2}\left(\frac{1}{N} \sum_{i=1}^{N} \frac{u_i}{U}\right)^2 + \frac{1}{2}\left(\frac{1}{N} \sum_{i=1}^{N} \frac{v_i}{U}\right)^2\right]$$

$$\approx U + \frac{1}{N} \sum_{i=1}^{N} u_i + \frac{1}{2N^2 U} \sum_{i=1}^{N} \left(u_i^2 + v_i^2\right) + \frac{1}{2N^2 U} \sum_{i=1}^{N} \sum_{j \neq i=1}^{N} \left(u_i u_j + v_i v_j\right)$$

(6.54)

This gives

$$V_{cup} \approx V_{SODAR} + \frac{1}{2U}\left[\frac{1}{N} \sum_{i=1}^{N} \left(u_i^2 + v_i^2\right)\right]$$

(6.55)

for large N. So $V_{cup} > V_{SODAR}$. Panofsky et al. (1977) show that

$$\frac{1}{N} \sum_{i=1}^{N} \left(u_i^2 + v_i^2\right) \approx \left(2.5u_*\right)^2$$

where is the friction velocity, and assuming a log wind profile

$$U = \frac{u_*}{\kappa_m} \ln \frac{z}{z_0}$$

where $\kappa_m \approx 0.4$ is the von Karman constant and z_0 is the roughness length. Since $V_{SODAR} \approx U$,

$$V_{cup} \approx V_{SODAR} \left\{ 1 + \frac{1}{2} \left[\ln \left(z / z_0 \right) \right]^{-2} \right\}$$

(6.56)

For example, over pasture having $z_0 = 0.05$ m, the correction is 1% at 50 m height. For rougher terrain or greater heights, the correction is smaller. The results of comparison between field trials and this theory have been inconclusive, possibly because of the limited height range and atmospheric conditions under which a log wind profile usually is observed (Antoniou and Jørgensen, 2003).

6.7.4 Calculating Wind Components from Incomplete Beam Data

In the presence of noise, one or more beams may have missing data at some range gate. "Missing data" are generally defined in some way by the SODAR manufacturer in terms of software switches which select various "filters" or consistency checks. This means that the usual matrix inversions may not be able to be used.

Two questions arise:

1. How should a reduced set of equations be solved to obtain estimates of wind components u, v, and w?
2. What is the effect on the uncertainties in an averaged wind when reduced data are used?

These questions relate to both calibration and operational issues because of the need to obtain the best possible data from a SODAR.

As described in Chapter 5 on calibration,

$$\hat{V} = \left(\hat{D}^T \hat{D} \right)^{-1} \hat{D}^T N$$

or, in the case of no instrument rotation

$$\hat{V} = \left(\hat{B}^T \hat{B} \right)^{-1} \hat{B}^T N$$

(6.57)

From this we obtain

$$\sigma_u^2 = \sum_{i=1}^{5} \left(\frac{\partial u}{\partial \eta_i} \right)^2 \sigma_\eta^2$$

for each of u, v, and w, where we only consider the 3-beam and 5-beam monostatic cases described in Chapter 5. Also, it is assumed that the same error in spectral peak position is present for all spectra. Missing data are accounted for by putting a zero into the B-matrix at that location. Six different standard deviations result. From the smallest to the largest,

$$\sigma_A = \frac{c}{2\sqrt{2}\, f_T \cos\theta}\, \sigma_{\Delta f}$$

$$\sigma_B = \frac{c}{2 f_T}\, \sigma_{\Delta f}$$

$$\sigma_C = \frac{c}{2\sqrt{2}\, f_T \sin\theta}\, \sigma_{\Delta f}$$

$$\sigma_D = \frac{c}{2 f_T \sin\theta}\, \sigma_{\Delta f}$$

$$\sigma_E = \frac{c\sqrt{3}}{2\sqrt{2}\, f_T \sin\theta}\, \sigma_{\Delta f}$$

$$\sigma_F = \frac{c\sqrt{1+\cos^2\theta}}{2 f_T \sin\theta}\, \sigma_{\Delta f}$$

$$(6.58)$$

giving the results in Table 6.2. Here $\eta_{ij} = (\eta_i - \eta_j)/2\sin\theta$, $p_{ij} = (\eta_i + \eta_j)/2\cos\theta$, $s_{ij} = (\eta_i \cos\theta - \eta_j)/\sin\theta$, $t_i = \eta_i/\sin\theta$.

6.7.5 WHICH GIVES LESS UNCERTAINTY: A 3-BEAM OR A 5-BEAM SYSTEM?

It is evident from the above that many combinations may occur in practice, with differing **error** contributions to the final averaged wind. The 5-beam system is more robust in terms of providing some measure of all three wind components, but acquisition of five beams takes 5/3 times as long as acquisition of three beams, so there will generally be 5/3 times as many wind estimates obtained at each range gate for a 3-beam system, giving a nominal $\sqrt{5/3} = 1.3$ times improvement in SNR.

Assume, because of noise, a random fraction β of spectra at a particular range gate produce acceptable data. For a 3-beam system the probability of obtaining acceptable spectra from beams 1 and 2, and thereby obtaining a wind speed estimate, is β^2. For a 5-beam system, the probability of obtaining acceptable data from beams 1 and 2, or 1 and 5, or 2 and 4, or 4 and 5 is

$$\beta^5 + 5\beta^4(1-\beta) + 7\beta^3(1-\beta)^2 + 4\beta^2(1-\beta)^3 = \beta^2\left(4 - 5\beta + 3\beta^2 - \beta^3\right)$$

The different terms on the left correspond to the probabilities of the acceptable combinations when five beams, four beams, three beams, and two beams have acceptable data. Overall, the ratio of acceptable 5-beam wind speeds to acceptable 3-beam wind speeds will be

$$\frac{3}{5}\left(4 - 5\beta + 3\beta^2 - \beta^3\right)$$

$$(6.59)$$

This is unity at $\beta = 0.7$ and increases for smaller f. This implies that a 3-beam system will generally give better quality data when data availability is higher (e.g., closer to the ground), but worse when data availability is reduced (e.g., further above the ground) as shown in Figure 6.13.

In addition, if full 3-beam data or full 5-beam data are available (the SNR is high), then Table 6.2 shows that the ratio of spectrum peak position errors is

$$\frac{\sigma_F}{\sigma_c} = \sqrt{1 + \cos^2\theta}\sqrt{2} \approx 2$$

(6.60)

TABLE 6.2

Formulae for computing velocity components from multi-beam SODARs when orientation $\phi = 0$ (for the definition of beam numbers refer to Fig. 4.42)

Beam	u	±	v	±	w	±
All	n_{14}	C	n_{25}	C	η_3	B
2345	s_{34}	F	n_{25}	C	η_3	B
1345	n_{14}	C	s_{35}	F	η_3	B
1245	n_{14}	C	n_{25}	C	$(p_{14}+p_{25})/2$	
1235	$-s_{31}$	F	n_{25}	C	η_3	B
1234	n_{14}	C	$-s_{32}$	F	η_3	B
123	$-s_{31}$	F	$-s_{32}$	F	η_3	B
124	n_{14}	C	$t_2-t_1-t_4$	E	p_{14}	A
125	$n_{12}+n_{15}$	E	n_{25}	C		A
134	n_{14}	C			η_3	B
135	$-s_{31}$	F	s_{35}	F	η_3	B
145	n_{14}	C	$n_{15}+n_{45}$	E	p_{14}	A
234	s_{34}	F	$-s_{32}$	F	η_3	B
235			n_{25}	C	η_3	B
245	$n_{24}+n_{54}$	E	n_{25}	C	p_{25}	A
12	t_1	D	t_2	D		
13	$-s_{31}$	F			η_3	B
14	n_{14}	C				
15	t_1	D	$-t_5$	D		
23			$-s_{32}$	F	η_3	B
24	$-t_4$	D	t_2	D		
25			n_{25}	C		
34	s_{34}	F			η_3	B
35			s_{35}	F	η_3	B
45	$-t_4$	D	$-t_5$	D		B

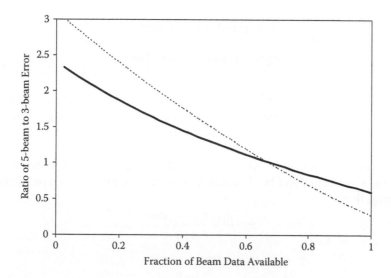

FIGURE 6.13 Estimated wind speed errors from a 5-beam system compared to a 3-beam system, as a function of fraction of individual spectra acceptable. Solid curve: neglecting peak position error dependence on SNR. Dashed curve: including peak position error.

so that a 5-beam system will be more accurate in this regime. A much more complex analysis using the probability of success of each beam combination times the estimated peak error gives the second curve in Figure 6.13, but the conclusion is effectively not changed.

6.8 SPATIAL AND TEMPORAL SEPARATION OF SAMPLING VOLUMES

The SODAR estimates wind components u, v, and w from at least three separated volumes. For example, at 100 m the data from u and v are separated by typically 40 to 50 m and 1.5 to 7 s (depending on the overall range). Assuming Taylor's frozen field hypothesis, the times 1.5 to 7 s correspond to distances of wind travel of 15 to 70 m at 10 m s^{-1}. The question arises as to how well correlated wind components are over these times and distances.

However, the distances characteristic of the SODAR operation are comparable or less than those applying in practice when a SODAR is used in conjunction with a wind turbine. Also, if calibrations are carried out against a mast, then such distances are also involved between the SODAR and mast. For calibration purposes, any fluctuations due to spatial and temporal separations will appear as added variance and uncertainty in fitted parameters.

Antoniou and Jørgensen (2003) and Antoniou et al. (2004) have shown that the distance between mast and SODAR is not a concern for a site which is on flat terrain.

For a 3-beam SODAR, assuming the vertical velocity is $w = 0$, the radial velocities recorded from tilted beams 1 and 2 are

$$\eta_1 = (u\cos\phi + v\sin\phi)\sin\theta$$

$$\eta_2 = (-u'\sin\phi + v'\cos\phi)\sin\theta \tag{6.61}$$

where θ is the beam tilt angle and ϕ is the SODAR orientation angle with respect to north (see Fig. 6.14). Because of spatial separation of the sampling volumes, the velocity components measured from each beam will not in general be exactly the same for a particular profile.

The components u and v are required, and also velocity components are averaged over a number of profiles. If solution for u and v is done on each profile *before* averaging, then

$$\hat{u} = \frac{\eta_1 \cos\phi - \eta_2 \sin\phi}{\sin\theta}$$

$$\hat{v} = \frac{\eta_1 \sin\phi + \eta_2 \cos\phi}{\sin\theta} \tag{6.62}$$

are the estimated components for each profile. This gives, from (6.61) and (6.62)

$$\hat{u} = u\cos^2\phi + u'\sin^2\phi$$

$$\hat{v} = v'\cos^2\phi + v\sin^2\phi \tag{6.63}$$

The square of the overall wind speed estimated from a single profile would be

$$\hat{V}^2 = \hat{u}^2 + \hat{v}^2 = (u^2 + v'^2)\cos^4\phi + 2(uu' + vv')\sin^2\phi\cos^2\phi + (u'^2 + v^2)\sin^4\phi$$

If this is now averaged,

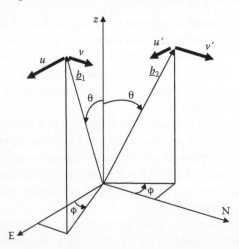

FIGURE 6.14 Geometry for beams 1 and 2, showing different measured wind components.

$$\overline{\hat{V}^2} = \overline{u^2 + v'^2}\cos^4\phi + 2\left(\overline{uu'} + \overline{vv'}\right)\sin^2\phi\cos^2\phi + \overline{u'^2 + v^2}\sin^4\phi$$

$$\overline{\hat{V}^2} \approx \overline{V^2}\cos^4\phi + 2\left(\overline{uu'} + \overline{vv'}\right)\sin^2\phi\cos^2\phi + \overline{V^2}\sin^4\phi$$

where $\overline{V^2}$ is the average of the square of the true wind. It is assumed that the average winds at the two sampling volume positions are the same, but that winds at both
locations are fluctuating (in the case of complex terrain however, then mean winds
at the two locations may be different). The terms $\overline{uu'}$ and $\overline{vv'}$ represent cross-cor-
relations between wind components at each sampling volume. We will assume they
can be written as

$$\overline{uu'} = \rho\overline{u^2}$$

$$\overline{vv'} = \rho\overline{v^2} \tag{6.64}$$

where ρ is a correlation coefficient. Also we assume $\overline{uu'} + \overline{vv'} \approx \rho\overline{V^2}$. The result is

$$\overline{\hat{V}^2} \approx \overline{V^2}\left[\cos^4\phi + 2\rho\sin^2\phi\cos^2\phi + \sin^4\phi\right] = \overline{V^2}\left[1 - (1-\rho)\sin^2 2\phi\right] \tag{6.65}$$

The estimated wind speed can be expected to be increasingly smaller than
the actual wind speed as the two sample volumes become more separated and ρ
decreases. This would be expected to give a lower correlation slope with mast measurements with increasing height.

If, as with the AeroVironment and Metek SODARs, the radial components or
spectra are averaged, then post-processing gives

$$\overline{\hat{u}} = \overline{u}\cos^2\phi + \overline{u'}\sin^2\phi = \overline{u}$$

$$\overline{\hat{v}} = \overline{v'}\cos^2\phi + \overline{v}\sin^2\phi = \overline{v}$$

so that there will be no beam-separation effect on wind speed estimates.

For a 5-beam SODAR, there are two extra beams, 4 and 5, tilted in the opposite
directions to beams 1 and 2. Similar to above

$$\hat{u} = \frac{\left(v_{r1} - v_{r4}\right)\cos\phi - \left(v_{r2} - v_{r5}\right)\sin\phi}{\sin\theta} = \frac{u + u'}{2}$$

$$\hat{v} = \frac{\left(v_{r1} - v_{r4}\right)\sin\phi - \left(v_{r2} - v_{r5}\right)\cos\phi}{\sin\theta} = \frac{v + v'}{2} \tag{6.66}$$

with the difference that there is no SODAR orientation dependence. Again there is a
beam separation effect *only* if the wind speed estimates are obtained *before* averaging. Then

$$\hat{V}^2 = \hat{u}^2 + \hat{v}^2 = \frac{1}{4}\left[V^2 + 2\left(uu' + vv'\right) + V'^2\right]$$

$$\overline{\hat{V}^2} = \overline{V^2}\left[1 - \frac{1}{2}(1-\rho)\right]$$

(6.67)

giving a similar expected decrease in slope with height, but without the orientation effect.

6.9   SOURCES OF MEASUREMENT ERROR

As with any instrumentation, poor maintenance, bad installation, and lack of understanding of operating principles can mitigate against good performance. Problems of this type are dealt with below, but in practice normal best practice will always enable quality measurements to be achieved with SODARs. For example, it should be possible to easily achieve wind speeds accurate to within a few percent over the entire profile of a well set-up SODAR.

6.9.1   HEIGHT ESTIMATION ERRORS

SODARs receive a continuous time record of scattered energy following each pulse transmission. Height z corresponding to the time t in the echo record is estimated based on the "adiabatic" speed of sound, c, which is the speed of sound in still air. Generally, the air temperature is measured at the instrument, so the sound speed c_0 is well-specified there. Also, a simple standard atmosphere change of temperature with height (this gradient is known as the "lapse rate") will give a next order approximation to the sound speed profile, although manufacturers probably do not do this. The sound speed, for a constant lapse rate at echo return time t, is

$$2\frac{dz}{dt} = c \approx c_0 + \frac{1}{2}\frac{c_0}{T_0}\frac{dT}{dz}z$$

(6.68)

giving

$$\ln\left(1 + \frac{1}{2T_0}\frac{dT}{dz}z\right) = \frac{1}{2T_0}\frac{dT}{dz}\frac{c_0}{2}t$$

If the height is estimated from $\hat{z} = c_0 t / 2$ then, in terms of the actual echo height z,

$$z = \frac{ct}{2} = \frac{c_0 t}{2} + \frac{1}{4}\frac{c_0 t}{T_0}\frac{dT}{dz}z = \hat{z}\left(1 + \frac{1}{2T_0}\frac{dT}{dz}z\right)$$

$$\hat{z} \approx z\left(1 - \frac{1}{2T_0}\frac{dT}{dz}z\right)$$

(6.69)

For example, dT/dz gives a negligible 0.3 m error at a range of 200 m. In comparison, a 15 K error in surface air temperature used to estimate c_0 gives a 5 m height error over the range 200 m. This error is readily corrected by measuring, or even estimating climatologically, the surface air temperature. If it is assumed that the return time for an echo from 200 m with a beam tilted at 18° is the same as that for a vertical beam, then the height error is 200(1−cos 18°) = 9.8 m. This error is also easily, although not necessarily, corrected by all SODAR manufacturers.

6.9.2 ERRORS IN BEAM ANGLE

The beam angle θ is vitally important in connecting the Doppler-shifted spectral peaks to the individual wind velocity components. This error has been discussed in Chapter 5, in which it was found, for $w = 0$, and an error $\Delta\theta$ in estimated beam zenith angle,

$$\hat{u} = \frac{\sin(\theta + \Delta\theta)}{\sin\theta} u \approx \left(1 + \frac{\Delta\theta}{\tan\theta}\right) u$$

(6.70)

with a similar expression for $\hat{v}$.

For a typical beam tilt angle of 18°, this represents a 5% error in wind speed for each 1° error in tilt angle. This order of error is unacceptable for wind energy applications which would normally require errors to be less than 1% overall.

The bias in effective beam angle due to a volume and power weighted $\sin\phi$ value has been discussed in Chapter 5. Other errors could occur if the acoustic baffle affects the overall shape of the beam and if this is not accounted for.

For dish antennas, the beam-pointing angle is determined by the alignment of the speaker/microphone units above the dish. For instruments such as the AQ500 in which the three beams are formed within a single solid structure, only misalignment due to some structural damage or change in the instrument could cause a $\Delta\theta$ change. However, for separated antennas, as with the Atmospheric Research Pty. Ltd. SODAR, setting up each antenna will be important.

For phased-array antennas, it might be at first thought that changes in the speed of sound due to temperature changes might alter the pointing angle. However, the pointing angle is actually determined by a time step Δt applied between speakers

$$kd \sin\theta = 2\pi f_T \Delta t$$

or

$$\sin\theta = \frac{f_T \lambda}{d} \Delta t = \frac{c_0 \Delta t}{d}$$

where c_0 is the speed of sound at the antenna.

For example, with the 3-beam monostatic SODAR, this gives

$$-\frac{c}{2f_T}\Delta f_1 = \frac{c_0 \Delta t}{d}u - \frac{c}{2f_T}\Delta f_3 \cos\theta$$

$$-\frac{c}{2f_T}\Delta f_2 = \frac{c_0 \Delta t}{d}v - \frac{c}{2f_T}\Delta f_3 \cos\theta$$

$$-\frac{c}{2f_T}\Delta f_3 = w$$

where c is the speed of sound at height z. This means

$$u = \left(1 + \frac{1}{2T_0}\frac{dT}{dz}z\right)\frac{d}{2f_T \Delta t}\left(-\Delta f_1 + \Delta f_3 \cos\theta\right)$$

$$v = \left(1 + \frac{1}{2T_0}\frac{dT}{dz}z\right)\frac{d}{2f_T \Delta t}\left(-\Delta f_2 + \Delta f_3 \cos\theta\right)$$

$$w = -\left(1 + \frac{1}{2T_0}\frac{dT}{dz}z\right)\frac{c_0}{2f_T}\Delta f_3$$

(6.71)

As an example, dT/dz gives only a negligible 0.3% error in estimated speed at z = 200 m if the lapse rate is not allowed for.

6.9.3 OUT-OF-LEVEL ERRORS

These errors in wind speed estimation were considered in Chapter 5, and are not significant for typical leveling errors.

6.9.4 BIAS DUE TO BEAM SPREAD

This **error** was also considered in Chapter 5. The error in assuming the central axis of a slanted beam defines the pointing angle is

$$m - 1 \approx \frac{4}{\left(ka\right)^2 \tan^2 \theta_0}$$

which, if $a = 0.4$ m, $k = 60$ m^{-1}, and $\theta_0 = 0.3$, yields a 7.3% **error** in overall wind speed This is by far the largest **error** likely to occur for SODARs.

6.9.5 BEAM DRIFT EFFECTS

The Doppler shift derives from reflections from a moving target in a moving medium traveling at an angle to the receiver. The situation can be visualized as the acoustic pulse being blown downwind. Scattered sound must then be directed further upwind in order to reach and be received by the SODAR. Both the upward and downward wind refraction effects cause extra Doppler effects and lead to a correction to the simple Doppler shift formula.

This was considered extremely carefully in Chapter 4. The result is that, instead of $N = BRV$ (assuming the tilt matrix $T = I$), we have

$$N = BRV + E \tag{6.72}$$

where

$$E = \frac{V^T V}{c} \begin{pmatrix} 1 \\ 1 \\ 1 \end{pmatrix} \tag{6.73}$$

for a 3-beam monostatic SODAR. The corrections are <1% for $V = 15$ m s^{-1}. If, on the other hand, the vertical radial wind information is not used, and $w = 0$, then the corrections are generally much larger, with

$$V \approx \hat{V} \left(1 + \frac{\hat{u} + \hat{v}}{c \sin \theta} \right) \tag{6.74}$$

as shown in Figure 6.15.

Corrected wind directions can be found similarly.

For a 5-beam monostatic SODAR system, winds are estimated from

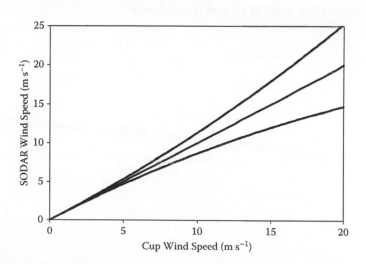

FIGURE 6.15 Wind speeds estimated for a 2-beam SODAR versus those which would be recorded by a cup anemometer. The upper and lower curves are the two extreme limits and the straight line shows 1:1.

$$u = \hat{u} = \frac{\left(\eta_1 - \eta_4\right)\cos\phi - \left(\eta_2 - \eta_5\right)\sin\phi}{2\sin\theta}$$

$$v = \hat{v} = \frac{\left(\eta_1 - \eta_4\right)\sin\phi + \left(\eta_2 - \eta_5\right)\cos\phi}{2\sin\theta}$$

$$w = \hat{w} + \frac{u^2 + v^2 + w^2}{c}$$

$$(6.74)$$

so the errors cancel out for the horizontal wind components, although remain for the vertical wind.

The error in the vertical velocity component, w, is about 0.3 m s^{-1} for all beam configurations when $V = 10$ m s^{-1}, so it can be a major source of error (Table 6.3). Also, there is error propagation into σ_w via

$$\sigma_w^2 = \sigma_{\hat{w}}^2 + \left(\frac{2V}{c}\right)^2 \sigma_V^2$$

$$(6.75)$$

Estimates of σ_w will therefore be too low.

As a simple test of this theory, the radial wind components from opposing beams in the 5-beam Metek were compared. The regression shown in Figure 6.16 is between the opposing beam 1 and beam 4 radial velocity components η_1 and η_4.

TABLE 6.3

Summary of beam-drift errors, assuming tilt angle $\theta = \pi/10$. (For errors in σ_w it is assumed $\sigma_w = 0.1$ m s^{-1} and $\sigma_V = 0.5$ m s^{-1}

System	Correction term	Maximum error at $V =$ 5 m s^{-1}	Maximum error at $V =$ 10 m s^{-1}	Maximum error at $V =$ 20 m s^{-1}
2-beam	$V \approx \hat{V}\left(1 + \frac{\hat{u} + \hat{v}}{c\sin\theta}\right)$	±9%	±18%	±36%
3-beam	$V \approx \hat{V}\left[1 + \frac{\left(\hat{u} + \hat{v}\right)\tan\frac{\theta}{2} + \hat{w}}{c}\right]$	±0.2%	±0.5%	±0.9%
5-beam		0	0	0
w	$w = \hat{w} + \frac{\hat{V}^2}{c}$	0.07 m s^{-1}	0.3 m s^{-1}	1 m s^{-1}
σ_w	$\sigma_{\hat{w}} = \sigma_w\left[1 - \frac{2V^2}{C^2}\frac{\sigma_V^2}{\sigma_w^2}\right]$	−1%	−4%	−17%

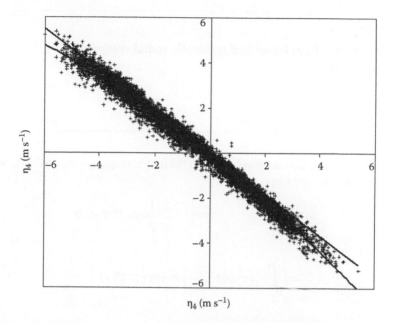

$\eta_4\ (\text{m s}^{-1})$

FIGURE 6.16 Radial velocities from beam 1 and beam 4 of the Metek SODAR at 28 m.

With exclusion of beam drift, $\eta_1 = \eta_4 + 2w\cos\theta$, and the inclusion of beam drift gives

$$\eta_1 = -\eta_4 + 2w\cos\theta - 2\frac{V^2}{c}\ .$$

As a test, we have done regressions of the linear fit $\eta_1 = a\eta_4$ and the quadratic fit $\eta_1 = a\eta_4 + b\eta_4^2$, in both cases assuming w to be negligible. The quadratic fit here makes the assumption that η_4 is on average proportional of the wind strength.

The results are shown in Table 6.4. The b coefficient can be a maximum of

$$\frac{2}{c\sin^2\theta} \sim 0.06\,\text{m}^{-1}\text{s}\ ,$$

but on average one could expect $\overline{u^2} = \overline{v^2} = \frac{1}{2}\overline{V^2}$ giving an expected b value of -0.03 as observed. Addition of the extra term in the regression does not explain much variance because the nonlinearity is small.

6.10   A MODEL FOR SODAR RESPONSE TO A PRESCRIBED ATMOSPHERE

Given the extensive background above, it is possible to generate a simple simulation code so that the SODAR response to a model atmosphere can be visualized. This is potentially useful for a transponder-based calibrator as well as for testing new design ideas.

TABLE 6.4

Parameters from linear and quadratic radial component regressions

	r^2	a	b
Linear	0.973	−0.92	
Quadratic	0.978	−0.97	−0.03

At time t echo signals are received from the height range $z_1 = c(t-\tau)/2$ to $z_2 = ct/2$. The echo signal has the form

$$p(t) = \int_{z_1}^{z_2} \frac{e^{-2\alpha z}}{z} \sigma_s(z) m\left(t - \frac{2z}{c}\right) \exp\left(j2\pi f_D t\right) dz$$

$$= \left[\int_{z_1}^{z_2} \sigma_s^*(z) m\left(t - \frac{2z}{c}\right) dz\right] \exp\left(j2\pi f_D t\right)$$

where the small lengthening or contraction of the pulse envelope m has been ignored, and the spherical spreading and absorption have been incorporated into $\sigma_s^*(z)$. The term $t-2z/c$ varies between 0 and τ over the range of the integral. Taking m as a Gaussian envelope,

$$m(t) = \exp\left[-\frac{1}{2\sigma_m^2}\left(t - \frac{\tau}{2}\right)^2\right]$$

$$= \exp\left[-\frac{1}{2\sigma_m^2}\left(\frac{2z}{c} - \frac{2}{c}\frac{z_1+z_2}{2}\right)^2\right]$$

$$= \exp\left[-\frac{4}{2\sigma_m^2 c^2}\left(z - \frac{z_1+z_2}{2}\right)^2\right]$$

or

$$m(z) = \exp\left[-\frac{1}{2\sigma_z^2}\left(z - \bar{z}\right)^2\right] \tag{6.76}$$

with $\sigma_z = \sigma_m c/2$. Then

$$p(t) = p_0(t)\exp\left(j2\pi f_D t\right) \tag{6.77}$$

where

$$p_0(t) \approx \int_{-\infty}^{\infty} \sigma_s^*(z) \exp\left[-\frac{1}{2\sigma_z^2}(z-\bar{z})^2\right] dz$$

(6.78)

The echo signal is sinusoidal at frequency f_D and with amplitude p_0 which is the Fourier transform, at wavenumber κ_x, of the Gaussian-weighted scattering cross-section one pulse length around height $c(t-\tau/2)/2$.

The power spectrum, without noise, has a Gaussian form centered on f_D and of width σ_f

$$P_k = \frac{P_0}{\sqrt{2\pi}\sigma_f} \exp\left[-\frac{1}{2\sigma_f^2}(f_k - f_D)^2\right] \qquad k = -\frac{n_s}{2}+1, -\frac{n_s}{2}+2, \ldots, \frac{n_s}{2}$$

The additive spectral noise *amplitudes* have a random Gaussian probability distribution of zero mean and standard deviation σ_N. Random deviates from the power spectrum with noise can be formed from

$$P_k - N\ln(U_1)$$

where U_1 is a random number uniformly distributed in the range 0 to 1.

6.11 SUMMARY

One of the key challenges for acoustic remote sensing is in obtaining robust estimates of physical parameters from signals accompanied by substantive noise. In this chapter we consider a wide range of approaches relating to this challenge. These include

1. Obtaining high quality estimates of the position of the peak in a noisy Doppler spectrum
2. Estimation of variances such as the variance of vertical velocity
3. Averaging methods within data of variable quality
4. Optimisation of multi-beam data when data from some beams are absent
5. Geometric errors due to pointing
6. Non-linear Doppler effects

REFERENCES

Antoniou I, Jorgensen H et al. (2004) Comparison of wind speed and power curve measurements using a cup anemometer, a LIDAR, and a SODAR. EWEC Wind Energy Conference, London.

Antoniou I, Jørgensen HE (2003) Comparing SODAR to cup anemometer measurements. EWEA Wind Energy Conference and Exhibition, Madrid.

Asimakopoulos DN, Helmis CG et al. (2002) Determination of the mixing height with an acoustic radar profilometer. In Argentini S, Mastrantonio G, Petenko I (Eds.), Proceedings of the 11th International Symposium on Acoust. Remote Sensing, Rome.

Bradley SG, Hünerbein SV (2004) A systematic algorithm for obtaining SODAR wind profiles. Meteorol Atmos Phys 85(1–3): 77–84.

Neff WD (1988) Observations of complex terrain flows using acoustic sounders: echo interpretation. Boundary Layer Meteorol 42(3): 207–228.

Panofsky HA, Tennekes H et al. (1977) The characteristics of turbulent components in the surface layer under convective situations. Boundary Layer Meteorol 11: 355–361.

Weill A, Klapisz C et al. (1980) Measuring heat-flux and structure functions of temperature-fluctuations with an acoustic Doppler Sodar. J Appl Meteorol 19(2): 199–205.

7 RASS Systems

Radio acoustic sounding systems (RASSs) are remote-sensing systems for the measurement of the temperature profile in the lower atmosphere. RASSs are deployed routinely in experiments and at monitoring sites as a simple addition to either a SODAR (SOund Detection And Ranging) or a RADAR windprofiler.

The essential feature of a RASS system is that it has an acoustic *transmitter* and a RADAR transmitter–receiver. The electromagnetic (EM) energy is reflected by the periodic refractive index variations created by the compressions and expansions of the air within the acoustic pulse. The RADAR wavelength is chosen to be half the acoustic wavelength so that EM reflections from successive acoustic compressions will combine in phase, giving a strong RADAR signal. By monitoring the acoustic properties, the speed of sound is deduced and hence the temperature.

One type of RASS, the *Doppler-RASS*, tracks an acoustic pulse with continuous EM waves. The Doppler effect provides a frequency shift which is used to determine sound speed and hence air temperature. Because of the continuous nature of the tracking wave, the EM transmitter and receiver are separate units.

An alternative design uses a continuous acoustic wave together with EM pulses. The echo is strongest when the acoustic and EM waves match the Bragg condition. The *Bragg-RASS* consists of an EM transmitter and acoustic transmitter and receiver units.

Other variants with continuous acoustic transmission and modulated EM transmission, or with both acoustic and EM pulsed transmissions, are also possible. A very good review is given by Kirtzel et al. (2000) (also see Vogt, 1966).

Because there is a combination of both acoustic and EM parameters here, we will use the subscript "a" for acoustic parameters and the subscript "e" for EM parameters. This means that the previous use of c for speed of sound is replaced by c_a in this chapter, and similarly for wavelength, frequency, and wavenumber.

7.1   RADAR FUNDAMENTALS

Historically RADAR was first used to track solid objects such as aircraft, and later precipitation was measured. Both these RADAR technologies rely on the measurements of the echo strength. When Doppler shift was first measured, wind speeds became accessible to measurements. Generally shorter wavelengths are used to obtain high reflectivity from hydrometeors and longer wavelengths to obtain high reflectivity from clear air refractive index changes. A good coverage of Doppler RADAR is given by Doviak and Zrnic (1984).

All RADARs, including the RASS-RADAR, use a stabilized local oscillator to generate a continuous signal which is modulated and amplified and fed to a klystron to produce a powerful microwave signal. Generally the transmitter is at the focus of a parabolic dish antenna so that a narrow beam is produced, and the receiver uses a comparable (or the same) dish antenna to provide a reasonable collecting area for scattered radiation and to focus the return signal onto a microwave receiver.

7.2   REFLECTION OF RADAR SIGNALS FROM SOUND WAVES

The power scattered back to a conventional RADAR from the atmosphere is described by a RADAR equation which is similar to the acoustic radar equation described in Chapter 4:

$$P_R = P_e G_e A_e \frac{c\tau}{2} \frac{e^{-2\alpha r}}{r^2} \sigma_s, \tag{7.1}$$

where P_e is the transmitted power, G_e the antenna transmitting efficiency into a solid angle, A_e the effective receiving area, $c\tau$ the length of the pulse in the atmosphere, r the range (generally taken to be the height z), α is an atmospheric absorption coefficient, and σ_s is the scattering cross-section.

However, when the acoustic source and the radar are (almost) collocated, and under the ideal conditions that the wavefronts of both the acoustic and radar waves are spherical with their center at the source point, the radar energy back-scattered from the acoustic wave will come to a focus at the radar set. This is in contrast to the r^{-2} one-way spreading loss associated with scattering from naturally occurring dielectric fluctuations of the atmosphere such as is associated with clear air turbulence.

This means that the equivalent RASS equation needs to follow a slightly different argument to find P_R. For example, the EM power incident on a Doppler-RASS acoustic pulse at range r is, for a RADAR half beam width $\Delta\theta$, $P_e G_e [\Delta\theta/4\pi]^2$ and the EM intensity is $P_e G_e [\Delta\theta/4\pi]^2 /4\pi r^2$. If the scattering cross-section per unit volume is σ_s, then the power scattered back to the RASS antenna is $P_e G_e \sigma_s N \lambda_a [\Delta\theta/4\pi]^2 /4\pi r^2$. The number of cycles in the acoustic pulse is N, so the length of the acoustic pulse is $N\lambda_a$.

The scattering cross-section, in general, includes Rayleigh scattering from precipitation particles (which has a λ_e^4 dependence) and scattering from atmospheric refractive index fluctuations. A structure function parameter C_n^2 for EM refractive index can be defined similarly to C_V^2 and C_T^2 in Chapter 2:

$$C_n^2 = \frac{\overline{[n(x)-n(0)]^2}}{x^{2/3}}. \tag{7.2}$$

The scattering cross-section per unit volume, σ_s, for refractive index changes has dimensions of m^{-1}. Physically, it can be expected to depend on C_n^2 (which has dimensions of m$^{-2/3}$) and the EM wavelength λ_e. A dimensional analysis gives

$$\sigma_s \propto C_n^2 \lambda_e^{-1/3}. \tag{7.3}$$

The proportionality constant is 0.38 (Hardy et al., 1966).

The refractive index of air at RADAR wavelength can be written as (Bean and Dutton, 1966)

$$n = 1 + \frac{77.6 \times 10^{-8}}{T}\left(P_{atm} + 4810\frac{e}{T}\right), \tag{7.4}$$

where p_{atm} is the air pressure in Pa, T the air temperature in K, and e the partial pressure of water vapor in Pa. Since typically $p_{atm} = 10^5$ Pa, $e = 10^3$ Pa, and $T = 280$ K, the moisture term is usually relatively minor.

If temperature fluctuations dominate, which is often the case for turbulence, and ignoring the moisture terms

$$\Delta n = -\frac{77.6 \times 10^{-8} p_{atm}}{T^2} \Delta T.$$

This provides a first-order connection between C_n^2 and C_T^2 as

$$C_n^2 \approx \left(\frac{77.6 \times 10^{-8} p_{atm}}{T^2} \right)^2 C_T^2. \tag{7.5}$$

There are three different mechanisms for scattering of EM radiation in clear air (Larsen and Röttger, 1991). Fresnel reflection is caused by a strong discontinuity of the refractive index perpendicular to the RADAR beam. Discontinuities in the atmospheric refractive index are usually in horizontal layers. With increasing zenith angle of the RADAR beam, the reflectivity due to horizontal layer discontinuities decreases rapidly. The relation between reflected power and elevation angle is called the *aspect ratio*. Within a scattering volume Fresnel scattering is caused by multiple discontinuities along the beam. For common RADARs, both Fresnel scattering and simple reflection are very small, which leaves Bragg scattering as the dominating mechanism.

Bragg scattering is caused by fluctuations of the refractive index having a spatial scale of $\lambda_e/2$. In the case of RASS instruments, the scattering is from an acoustic pulse, so the scattering cross-section is from the acoustic wave variations. These depend in amplitude on the transmitted acoustic power P_a, and have both pressure and temperature variations associated with them.

From (7.4) and ignoring moisture, we obtain

$$\frac{\Delta n}{n} = 77.6 \times 10^{-8} \left(\frac{\Delta p}{T} - \frac{p_{atm}}{T^2} \Delta T \right) \approx 2.7 \times 10^{-9} \Delta p - 9.8 \times 10^{-7} \Delta T,$$

where it is assumed a standard atmosphere pressure of $p_{atm} \approx 1.013 \times 10^5$ Pa and $T = 283$ K. Also, from Chapter 2, the hydrostatic equation gives $\Delta p = -\rho g \Delta z$ and the adiabatic lapse rate gives $\Delta T = -g \Delta z / c_p$ which, when combined, give $\Delta T = \Delta p / \rho c_p \approx 8 \times 10^{-4} \Delta p$. The net result for sound waves, which undergo adiabatic expansions and compressions, is

$$\frac{\Delta n}{n} \approx 2 \times 10^{-9} \Delta p. \tag{7.6}$$

For an acoustic wave, the amplitude of pressure variations Δp is related to the acoustic intensity I_a through $\Delta p = \sqrt{2 \rho c_a I_a}$. The acoustic intensity is just the acoustic power transmitted divided by the area at distance r, so

$$\frac{\Delta n}{n} \approx 2 \times 10^{-9} \sqrt{2\rho c_a I_a} = 2 \times 10^{-9} \sqrt{2\rho c_a \frac{G_a P_a}{4\pi r^2}} \approx \frac{1.6 \times 10^{-8}}{r} \sqrt{G_a P_a}.$$

The amplitude of scattered EM radiation depends on the refractive index variation Δn, and so the scattered intensity depends on $(\Delta n)^2$. In Chapter 2, it was found that interaction between a sinusoidal acoustic pulse and refractive index fluctuations gave a sinc function for amplitudes

$$\frac{\sin[(2k-\kappa)c\tau/2]}{(2k-\kappa)c\tau/2}.$$

For a Doppler-RASS, the length of the acoustic pulse is $c\tau = N\lambda_a$, the wavenumber k of the interrogating wave is k_e, and the spatial wavenumber of the fluctuations is k_a. The scattering cross-section therefore has the form

$$\sigma_s \propto \left\{ \frac{\sin[(2k_e - k_a)N\lambda_a/2]}{(2k_e - k_a)N\lambda_a/2} \right\}^2 \left[\frac{1.6 \times 10^{-8}}{r} \sqrt{G_a P_a} \right]^2 N\lambda_a (2r\Delta\theta)^2$$

$$\propto G_a P_a N\lambda_a \left\{ \frac{\sin[(2k_e - k_a)N\lambda_a/2]}{(2k_e - k_a)N\lambda_a/2} \right\}^2 (\Delta\theta)^2. \tag{7.7}$$

Note that this peaks sharply at the Bragg condition

$$2k_e = k_a, \qquad \lambda_e = 2\lambda_a. \tag{7.8}$$

The $N\lambda_a (2r\Delta\theta)^2$ term represents the volume illuminated at range r by a beam of half-beam-width $\Delta\theta$:

$$P_R \propto \left(\frac{\lambda_a}{\lambda_e}\right)^2 \frac{(\Delta\theta)^4 P_e G_e^4 P_a G_a N^2}{r^2} \left\{ \frac{\sin[(2k_e - k_a)N\lambda_a/2]}{(2k_e - k_a)N\lambda_a/2} \right\}^2 L(r). \tag{7.9}$$

The λ_e factor arises because the efficiency of an EM antenna depends on wavelength. The exponential absorption term has been replaced by $L(r)$ which represents losses due to scattering out of the beam and depends on C_n^2. This term determines the range limitation of the RASS. Clifford and Wang (1977) give a full derivation of P_R, which is an extension of the derivation by Marshall et al. (1972).

The dependence on the pulse length and the Bragg condition in (7.9) is of the form

$$\left\{ \frac{\sin[N(2(k_e/k_a)-1)\pi]}{(2(k_e/k_a)-1)\pi} \right\}^2.$$

This is plotted in Figure 7.1.

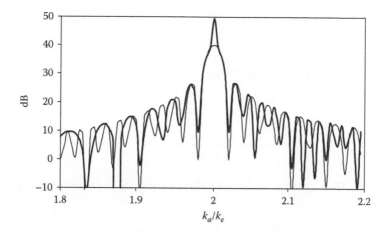

FIGURE 7.1 The sensitivity of received power to the Bragg condition for $N = 100$ (fine line) and $N = 300$ (dark line).

7.3 ESTIMATION OF MEASURED HEIGHT

The RASS unit sends out an acoustic wave in the vertical direction. The propagation speed of the acoustic wave depends on the temperature and moisture composition of the atmosphere. The following is based on the description provided by Metek for their DSDPA.90 SODAR/MERASS.

Given that the EM wave is continuous for a Doppler-RASS, the actual measurement height z_r is determined from the time t_a elapsed after the transmission of the acoustic pulse, as shown in Figure 7.2.

$$z_r = \int_0^{t_a - z/c_e} c_a(t)\,dt. \tag{7.10}$$

The average sound speed over this height range is given by

$$\overline{c}_a = \frac{\displaystyle\int_0^{t_a - z_r/c_e} c_a(t)\,dt}{t_a - z_r / c_e}. \tag{7.11}$$

From (7.10) and (7.11),

$$z_r = \overline{c}_a\left(t_a - \frac{z_r}{c_e}\right) = \frac{\overline{c}_a t_a}{1 + \overline{c}_a / c_e} \approx \overline{c}_a t_a. \tag{7.12}$$

To calculate $\overline{c}_a$, either (7.11) is used based on the RASS measurements or the sound speed derived from a nearby surface temperature (ideally also a humidity sensor) can be used.

From the frequency shift Δf of the reflected EM waves of wave number k_e, the local sound velocity c_a is derived from the Doppler equation

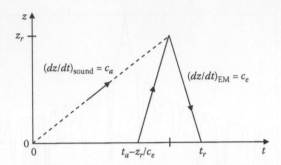

FIGURE 7.2   The timing of acoustic and EM signals propagating to and from height z_r.

$$\Delta f_e = -\frac{2c_a}{c_e}f_e = -\frac{2c_a}{\lambda_e} = -\frac{k_e c_a}{\pi}. \tag{7.13}$$

This sound speed also contains effects from humidity fluctuations and the wind speed along the beam. If the value of the vertical wind speed is larger than the measurement error, the sound speed can be corrected for this effect. However the vertical wind speed is usually very small.

7.4   DEDUCTION OF TEMPERATURE

7.4.1   DOPPLER-RASS

From Chapter 3, the speed of sound is related to the temperature by

$$c = \sqrt{\frac{\gamma_{\text{dry air}}RT}{M_{\text{dry air}}}\left[1 + \left(1 - \varepsilon - \frac{2}{35}\right)\frac{e}{p}\right]} \approx \sqrt{\gamma_{\text{dry air}}R_d T_v} \approx 20.05\sqrt{T_v}\ \text{m s}^{-1}. \tag{7.14}$$

Besides the second-order effects of humidity and vertical wind, there are some third-order variations caused by the ideal gas approximation, cross-wind influence, cross-wind/turbulence, and turbulence. Sound velocity is, from (7.13),

$$c_a = -\Delta f_e \frac{c_e}{2f_e}.$$

Typically, $f_e = 1290$ MHz, $c_e = 3 \times 10^8$ m s^{-1}, and $c_a \approx 340$ m s^{-1}, so $\Delta f_e \approx 3$ kHz. In the Metek RASS, the received signal is mixed with f_m and low-pass filtered to give an audio frequency signal, which is much easier to process. First the local air temperature T_s is measured at the surface and then the expected frequency shift Δf_s = $(\Delta f_e)_{\text{surface}}$ calculated from (7.13) for this surface value of sound speed c_s. Then the mixing frequency is set at $f_m = f_e + \Delta f_s$. The result of the mixing process is to produce a spectrum centered on $f_{beat} = f_e - f_m = -\Delta f_s$ (recall that, since the sound is moving away from the RASS, Δf_s is negative). At the surface, the spectrum will have a peak at 0 Hz. The sound speed is now calculated from

$$c_a = (f_{beat} + \Delta f) \frac{c_e}{2 f e},$$

(7.15)

where Δf is the first moment of the spectrum (the frequency shift of the spectral peak from the center of the spectrum). In practice, f_{beat} is forced to the nearest spectral estimation frequency, since this removes any initial systematic bias. Note that, since $f_e = 1290$ MHz is a frequency allocated to this type of instrument, the Bragg condition implies

$$f_a = \frac{k_a c_a}{2\pi} = \frac{2 k_e c_a}{2\pi} = 2 f_e \frac{c_a}{c_e}.$$

(7.16)

Based on $c_a \approx 340$ m s^{-1} and $c_e = 3 \times 10^8$ m s^{-1}, this gives $f_a = 2924$ Hz. Therefore an acoustic frequency of close to 3 kHz needs to be transmitted. A Doppler-RASS may also have modulation of the acoustic pulse to help obtain a Bragg condition match.

7.4.2 BRAGG-RASS

The Bragg-RASS uses a continuous acoustic wave and a pulsed EM signal. Consider an acoustic pressure peak at a height z at time t, as shown in Figure 7.3. At time λ_a/c_a this pressure peak has moved upward to height $z + \lambda_a$. Now the continuous acoustic wave looks exactly as it did at time t. This means that EM reflections from the acoustic wave will be identical at time t and at time $t + \lambda_a/c_a$. The variations in the amplitude of the scattered EM wave must therefore have a period of λ_a/c_a. This means that

$$\Delta f_e = \frac{c_a}{\lambda_a} = f_a.$$

(7.17)

The rather surprising result is that the Doppler shift equals the acoustic frequency and the Doppler shift provides no information on temperature structure. Instead, the

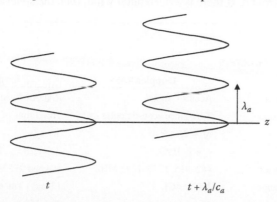

FIGURE 7.3 The time taken for identical reflected EM amplitude from the continuous acoustic wave in a Bragg-RASS.

change in sound speed is sensed by modulating the acoustic frequency or providing sufficient acoustic bandwidth so that the Bragg condition is bracketed by the range in f_a. Then the peak in the EM spectrum indicates the frequency at which

$$\Delta f_e \mid_{\max} = \frac{2c_a}{\lambda_e}. \tag{7.18}$$

7.5   WIND MEASUREMENTS

It is possible to use a RASS system to also measure wind profiles, in exactly the same manner as described for monostatic SODARs. Typically four tilted beams and one vertical beam are used for both acoustic transmission and EM scattering. The off-vertical beams introduce an extra Doppler shift corresponding to the radial velocity. The horizontal and vertical wind components can then be measured in analogy to the 5-beam SODAR principle.

7.6   TURBULENCE MEASUREMENTS

Sound speed fluctuations in the vertical direction are dominated by wind speed fluctuations even under convective conditions. The contribution of steady convective updrafts or downdrafts is about 10% in strongly convective conditions and can therefore be neglected. RASS therefore yields the *turbulent* vertical wind fluctuations (Kirtzel et al., 2000).

7.7   RASS DESIGNS

Table 7.1 summarizes typical parameters of the two RASS types (Engelbart, 1998).

Various physical layouts have been used. One of the problems to be addressed is that the sound spreads out from the acoustic source in a spherical wave. Reflection of the EM wave from the spherical acoustic wave focuses the scattered energy back toward the ground. If there is a horizontal wind, then the spherical wave moves

TABLE 7.1

Typical RASS parameters

	Doppler-RASS	Bragg-RASS
Frequency modulation	Acoustic signal	RADAR signal
Height z estimated from	Time since acoustic pulse t_a	Travel time of EM pulse t_e
Frequency shift	$\Delta f_e = 2f_e c_a/c_e$	$\Delta f_e = f_a$
Sound speed	$c_a = c_e \Delta f_e/2f_e$	$c_a = c_e \Delta f_e\mid_{max}/2f_e$
Typical EM frequencies	482, 915, 1270–1295 MHz	404 and 915 MHz
Typical maximum range	200 m AGL	13 and 1 km, respectively
Typical resolution	30 m	300, 150 m
Antenna diameters	1.5 m	12, 100 m

downwind and so does its focus. This means that the RASS gradually loses *extra* (compared to the normal spherical spreading and scattering losses) signal strength as the height increases (Lataitis, 1992). The situation is shown in Figure 7.4. The Metek RASS (Figure 7.5) uses the configuration shown in Figure 7.6.

A configuration using one EM transmitter or windprofiler and four acoustic antennae is shown in Figure 7.7. Wind direction determines which acoustic

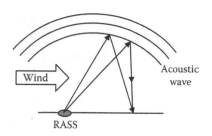

FIGURE 7.4 Movement of the focus downstream.

antenna serves as the acoustic transmitter (Angevine et al., 1994). Another approach is to have two EM profilers with one acoustic antenna and, depending on wind direction, the whole instrument can be rotated around its axis. The wind speed determines the distance between profiler and acoustic antenna (Vogt, 1966) as shown in Figure 7.8. In this system, both bistatic and monostatic configurations can be used, and both RADAR and SODAR can be continuous and/or pulsed. Also various kinds of frequency modulation can be applied to either RADAR or SODAR signals. Bistatic arrangements can overcome wind drift effects to a certain extent (improve the range by a factor of 4). Bistatic systems enable measurement of horizontal wind speed and direction by measuring delay times between the different antenna sites. Combinations of bistatic and monostatic configurations have been developed to overcome orientation problems of bistatic systems.

FIGURE 7.5 The Metek MERASS.

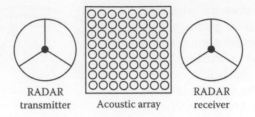

FIGURE 7.6   The layout of the Metek RASS.

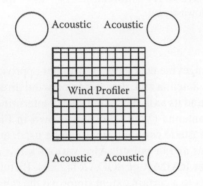

FIGURE 7.7   A RASS configuration which uses the best of four acoustic antennas, depending on the wind direction.

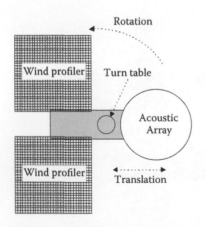

FIGURE 7.8   A turntable RASS with the axis of operation aligned with the wind.

7.8   ANTENNAS

The EM transmitter/receiver can either be a dedicated RADAR or an EM wind-profiler which is also obtaining wind information from back-scattered EM radiation (Skolnik, 2001; Klaus et al., 2002). For dedicated RADAR units, the transmitted dish and receiver dish are generally separated, as for the Metek unit in Figure 7.5. A common transmitter/receiver unit would struggle with overload problems. Dish separation distance is typically two to three aperture diameters (4–6 m for a 1290 MHz system). This configuration is strictly speaking bistatic, and scattering is not 180° and is also height-dependent. The Metek MERASS uses EM radiation "leaked" from the side of the transmitter unit, and received by the other dish, as a reference signal to beat with the signal scattered off the acoustic pulse, so as to form an audio frequency Doppler shift signal. A trailer-mounted Metek RASS system is shown in Figure 7.9. The receiving dish can be seen in the foreground and the edge of the transmitting dish at the other end of the trailer. The center of the trailer is occupied by a SODAR, with thnadners, which operates as a SODAR at 1750 Hz and as a RASS acoustics unit at 3 kHz. The units covered with white plastic in the foreground are PC and amplifier units. Beyond the trailer is a smaller Metek SODAR.

FIGURE 7.9 A Metek SODAR/RASS mounted on a trailer, together with a smaller Metek SODAR.

7.8.1 BAFFLES

EM screening is sometimes required for RASS units (Skolnik, 2001). For example, Figure 7.10 shows an Atmospheric Research Pty. Ltd. RASS (the surrounding fence is not normally required).

7.9 LIMITATIONS

See Angevine and Ecklund 1994.

FIGURE 7.10 Screening required between EM transmitter and receiver on an Atmospheric Research Pty. Ltd. RASS.

7.9.1   RANGE

The maximum altitude covered by a RASS system depends on its wavelength and meteorological conditions. Usually the range is determined by the drift of the acoustic wavefront with the horizontal wind. Consider the case time t after transmission of the acoustic pulse. The acoustic wave has moved distance ut downstream, where u is the wind speed. From symmetry in Figure 7.4, when a ray from the RASS reflects from the center of the curved acoustic wave, distance ut downstream, the reflected ray reaches the ground distance $2ut$ downstream. In the same time, the acoustic pulse has reached a height $z = c_a t$. The result is that the drift of the focus downstream, when the reflection is from a height z, is

$$D = 2 \int_0^z \frac{u(z)}{c_a(z)} \, dz. \tag{7.19}$$

When the focus leaves the receiving antenna aperture, the received power drops drastically. The focus is not perfect though, and atmospheric turbulence tends to spread out the acoustic pulse. The best atmospheric condition is buoyancy-driven turbulence at low wind speed, whereas the opposite is true for low turbulence at high wind speed. Only occasionally does RASS range reach as high as 1.5 km. The upper limit is imposed by the attenuation of the sound waves, distortion of the acoustic wave fronts by turbulence, and advection of the sound out of the RADAR beam. The vertical air velocity measurements are distorted by ground and intermittent clutter in the RADAR side lobes, mainly below 500 m. New signal-processing techniques have been developed by Potvin and Rogers (1999) to overcome these limitations. Specifically, a technique involving order statistics of several consecutive temperature and vertical velocity measurements, known as the median filter, helps to filter out most of the bad measurements and to estimate vertical heat flux. The elimination of bad measurements also allows the study of the development and structure of convection in the boundary layer. The RASS has a unique role in boundary layer studies since it can collect data at altitudes higher than most instrumented towers, and over periods of time longer than those obtained by instrumented airplanes.

Typically, RASS-derived temperature profiles reach up to the maximum height of the associated wind profiler, or less: for a 915-MHz system, that is 1 to 3 km above ground level (1).

7.9.2   TEMPERATURE

There are many studies in which the temperature errors of RASS have been investigated (e.g., see Angevine and Ecklund, 1994; Angevine et al., 1994; Petenko, 1999; Gorsdorf and Lehmann, 2000).

Figure 7.11 shows a correlation between mast measurements and the Metek MERASS as part of the WISE/PIE campaign (Bradley et al., 2004). The correlation is at 100 m, and comprises 3664 points. The RMS residual was 0.37 K.

The availability of RASS temperatures during a period of the same campaign is shown in Figure 7.12. The average height producing good data is about 110 m. This is considerably lower than the SODAR which uses the same acoustic antenna.

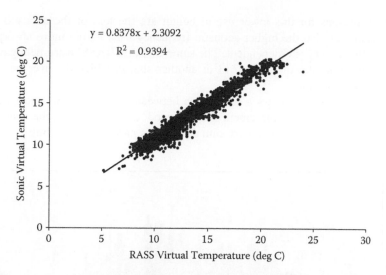

FIGURE 7.11   A correlation between Metek RASS temperatures and mast temperatures at 100-m height.

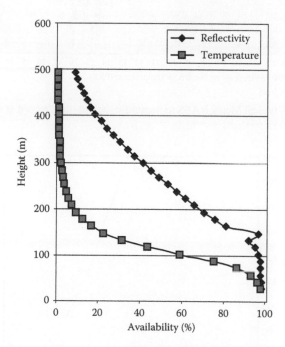

FIGURE 7.12   Data availability of the Metek RASS temperature data versus height at a site in Denmark.

The main reasons for this lower useful height are the loss of the acoustic wave downstream and also the higher acoustic frequency used giving more absorption. However, this is very site-dependent. The same SODAR/RASS instrument gave the temperature–time plot of Figure 7.13 at another site, in which the average useful height is above 300 m.

Finally, Figure 7.14 shows a correlation between a Metek RASS and temperatures from a radiosonde. Agreement is quite good, considering that the sonde will have drifted a long distance downstream during this part of the sounding.

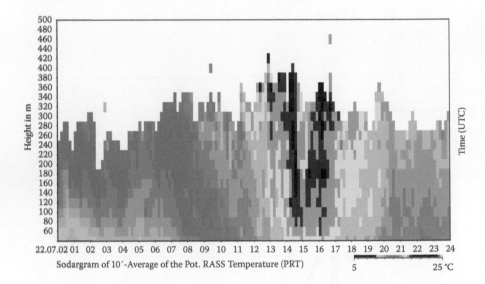

FIGURE 7.13   A typical Metek RASS temperature–height–time record at a UK site.

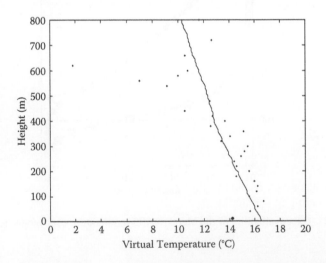

FIGURE 7.14   Comparison between a Metek RASS (dots) and a radiosonde (solid line).

7.10 SUMMARY

This chapter has given a very brief overview of the principles underlying RASS technology.

REFERENCES

Angevine WM, Ecklund WL (1994) Errors in radio acoustic sounding of temperature. J Atmos Ocean Technol 11: 837–842.

Angevine WM, Ecklund WL et al. (1994) Improved radio acoustic sounding techniques. J Atmos Ocean Technol 11: 42–49.

Bean BR, Dutton EJ (1966) Radio meteorology. US Government Printing Office, Washington, DC.

Bradley SG, Antoniou I et al. (2004) SODAR calibration for wind energy applications. Final reporting on WP3 EU WISE project NNE5-2001-297.

Clifford SF, Wang TI (1977) Range limitation on radar-acoustic sounding systems (RASS) due to atmospheric refractive turbulence. IEEE Trans Antennas Propagation 25(3): 319–326.

Doviak RJ, Zrnic DS (1984) Doppler radar and weather observations. San Diego, Academic Press, 562 pp.

Engelbart D (1998) Determination of boundary layer parameters using wind-profiler/RASS and SODAR/RASS. 4th International Symposium on Tropospheric Profiling, Sowmass, Colorado.

Gorsdorf U, Lehmann V (2000) Enhanced accuracy of RASS-measured temperatures due to an improved range correction. J Atmos Ocean Technol 17: 407–417.

Hardy KR, Atlas D et al. (1966) Multi-wavelength backscatter from the clear atmosphere. J Geophys Res Atmos 71(6): 1537–1552.

Kirtzel HJ, Voelz E et al. (2000) RASS – a new remote sensing system for the surveillance of meteorological dispersion. Kerntechnik 65(4): 144–151.

Klaus V, Cherel G et al. (2002) RASS developments on the VHF radar at CNRM/Toulouse height coverage optimization. J Atmos Ocean Technol 19: 967–979.

Larsen MF, Rottger J (1991) VHF RADAR measurements of in-beam incidence angles and associated vertical-beam radial-velocity corrections. J Atmos Ocean Technol 8: 477–490.

Lataitis RL (1992) Signal power for radio acoustic sounding of temperature: the effects of horizontal winds, turbulence, and vertical temperature gradients. Radio Sci 27: 369–385.

Marshall JM, Barnes AA et al. (1972) Combined radar-acoustic sounding system. Appl Opt 11(1): 108.

Petenko IV (1999) Improved estimation of errors due to antenna geometry in RASS based on a RADAR wind profiler. Met Atmos Phys 71: 69–79.

Potvin G, Rogers RR (1999) Measuring vertical heat flux with RASS. Met Atmos Phys 71: 91–103.

Skolnik MI (2001) Introduction to radar systems. New York, McGraw-Hill. 581 pp.

Vogt S (1966) Advances in RASS since 1990 and practical application of RASS to air pollution and ABL studies. 8th International Symposium on Acoustic Remote Sensing.

8 Applications

This book is primarily about the design and operating principles of atmospheric acoustic remote-sensing instruments, so this chapter will simply give a few examples of the use to which this technology can be put. For a more exhaustive insight into applications, there are very good review articles such as Singal (1997), Asimakopoulos (1994), Asimakopoulos and Helmis (1994), Asimakopoulos et al. (1996), Engelbart (1998), Reitebuch and Emeis (1998), Coulter and Kallistratova (1999), Engelbart et al. (1999), Helmis et al. (2000), Kirtzel et al. (2000), Melas et al. (2000), Ostashev and Wilson (2000), Seibert et al. (2000), Emeis (2001), Engelbart and Steinhagen (2001), Piringer and Baumann (2001), Raabe et al. (2001), Ruffieux and Stübi (2001), Neisser et al. (2002), Peters and Fischer (2002), Anderson (2003), and Bradley et al. (2004b).

8.1   REVIEW OF SELECTED APPLICATIONS

8.1.1   Environmental Research

A major use of SODAR and RASS technology is in monitoring and understanding the atmospheric boundary layer in relation to air pollution and dispersion modeling. Traditionally it has been difficult for these instruments to work effectively in closely built-up urban areas, because of echoes from buildings and because of impact on residents, but this is changing as the acoustic design of the instruments improves. We give here a few results from Salfex, an urban "street canyon" momentum and heat flux study in Salford, Greater Manchester, UK, which was led by Janet Barlow of Reading University (Barlow et al., 2007).

Figure 8.1 shows a site plan of the street canyon study area and the SODAR location. The SODAR was placed on the other side of the River Irwell, with relatively open land upwind to the north, but within 30 m of occupied housing to the east. Directly measuring instrumentation included masts extending to just above the dense housing in the study area, and the AeroVironment 4000 SODAR provided data above that height. In this way, wind profiles could be obtained at regular intervals, such as the half-hourly profiles shown in Figure 8.2.

FIGURE 8.1   Site plan for the Salfex campaign. The street canyon measurements were at site 1, and the SODAR at site 2. The plot is 1 km on each edge.

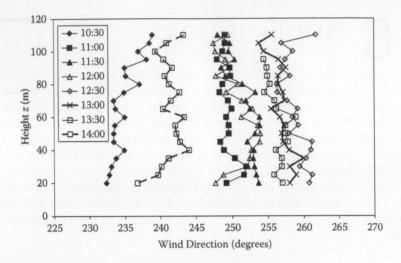

FIGURE 8.2 Wind direction profiles recorded every half hour.

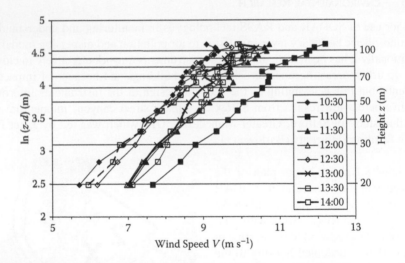

FIGURE 8.3 Logarithmic wind speed profiles measured every half hour.

Estimates of roughness length in the complex surface of the streets and buildings were readily available, as shown in the example of Figure 8.3.

The lowest points, at $z-d = 12$ m (with $d = 8$ m) represented the lowest height accessible to the SODAR (because of ringing within the baffle). The roughness length z_0, friction velocity u_*, and drag coefficient $(u_*/v)^2$ all show variation with wind direction. This is not surprising given the clearer sectors, but it would be difficult to quantify these variations with any other instrument than a SODAR.

Second-moment data, such as the results for $\sigma_{u,v}/\sigma_w$ shown in Figure 8.4, indicate a change in the boundary layer regime at about 80 m. It is the interpretation of this

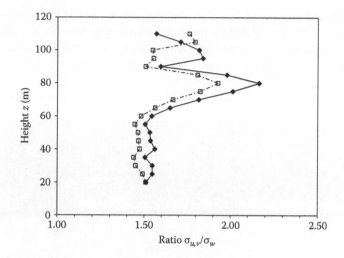

FIGURE 8.4 Second moment profiles. σ_u/σ_w (diamonds), σ_v / σ_w (squares).

type of observed feature which is particularly useful in guiding the development of new models for this challenging area of meteorology.

8.1.2 BOUNDARY LAYER RESEARCH

The use of an array of SODARs presents some interesting measurement opportunities. These include being able to investigate advection of non-turbulent structures. The SABLE SODAR array (Bradley et al., 2004b; Bradley and von Hunerbein, 2006) consisted of four vertically pointing speaker-dish units having individual power amplifiers and local intelligence. They were interconnected via RS485 operat-

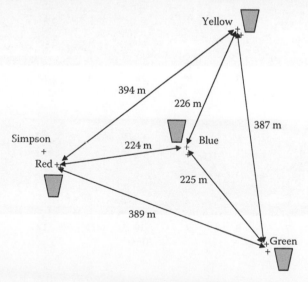

FIGURE 8.5 The geometry of the Antarctic SODAR array.

ing at 57.6 kB and exchanged data with a centralized PC. SODAR spacing was about 400 m (Figure 8.5). The SODARs transmitted simultaneously in non-overlapping frequency bands, but with center frequency, pulse characteristics, sampling, and other parameters selectable on a pulse-by-pulse basis. Local control was achieved with microprocessors. The array comprised three SODARs placed at the vertices of an equilateral triangle, and a fourth SODAR at the center. Figure 8.6 shows typical

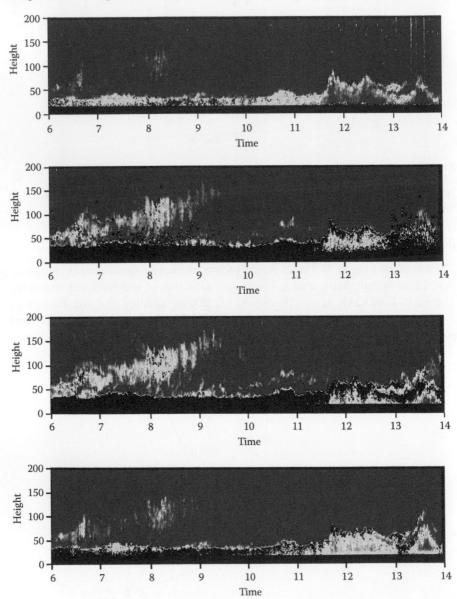

FIGURE 8.6 (See color insert following page 10). Plots of time variations of the C_T^2 field measured by the four SODARs.

time series of C_T^2 profiles. Fluctuations in C_T^2 occur at each range gate level, and these are often correlated across the four SODARs because of advected coherent structures. Covariances were computed at each height for each pair of SODARs and from these the corresponding time lags were estimated. This resulted in a system of linear equations to be solved for the advected velocity components (u, v), as follows.

$$
\hat{u}\begin{bmatrix} \Delta x_{y,r} \\ \Delta x_{g,r} \\ \Delta x_{g,y} \\ \Delta x_{b,r} \\ \Delta x_{b,y} \\ \Delta x_{b,g} \end{bmatrix} + \hat{v}\begin{bmatrix} \Delta y_{y,r} \\ \Delta y_{g,r} \\ \Delta y_{g,y} \\ \Delta y_{b,r} \\ \Delta y_{b,y} \\ \Delta y_{b,g} \end{bmatrix} = \begin{bmatrix} \Delta r_{y,r}^2 / \tau_{y,r} \\ \Delta r_{g,r}^2 / \tau_{g,r} \\ \Delta r_{g,y}^2 / \tau_{g,y} \\ \Delta r_{b,r}^2 / \tau_{b,r} \\ \Delta r_{b,y}^2 / \tau_{b,y} \\ \Delta r_{b,g}^2 / \tau_{b,g} \end{bmatrix},
$$

(8.1)

where the Δx and Δy are the components of the vector Δr between each pair of SODARs, and the τ values are the estimated time lags based on correlations at each range gate of pairs of C_T^2 versus time records. Figure 8.7 shows the matrix of covariances versus height, with obvious peaks at each height which can give the τ values. This method yields wind profiles from non-Doppler SODARs, as shown in Figure 8.8. The technique also allows for estimates of the size of coherent structures, based on the covariance matrix.

8.1.3 WIND POWER AND LOADING

We have already presented calibration data from the WISE project in previous chapters (Bradley et al., 2004a). The aim of that project was to prove that SODARs have sufficient reliability and accuracy for the rather demanding wind-power industry requirements (better than 1% accuracy at all heights to 150 m with high data availability). Figure 8.9 shows the field calibration layout.

From profiles produced by SODARs, it is possible to monitor turbine performance as a function of wind speed and to do this with considerable accuracy as shown in Figure 8.10 (Antoniou et al., 2004).

8.1.4 COMPLEX TERRAIN

SODAR and RASS are relatively portable devices and can operate from a small generator or battery-backed solar cells. This makes them a useful technology for investigations of flows and mixing layer heights in complex terrain. Most of the journal literature relating to acoustic remote sensing in the atmosphere describes such measurements.

Here we simply show some of the information which is available. First, Figure 8.11 shows wind profiles measured by an AeroVironment 4000 SODAR from prior to dawn through sunrise. Two aspects are very evident: the useful height range is greatly reduced during the night in this example, when turbulence is suppressed

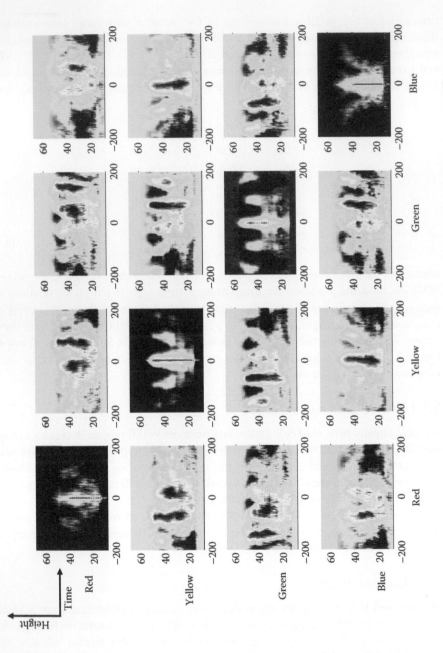

FIGURE 8.7   (See color insert following page 10). Matrix of covariances between C_T^2 values measured by each pair of SODARs at each height.

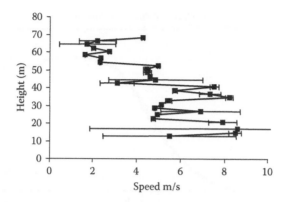

FIGURE 8.8   Wind profiles derived from the covariances shown in Figure 8.7.

FIGURE 8.9   The field layout for the WISE calibration campaign. From foreground to background: Scintec SODAR with large-diameter white baffles; AeroVironment 3000; Metek SODAR/RASS (with RADAR dish); and a small Metek SODAR.

because of the cool surface; and there are intriguing wind direction changes with height (but not significant change in wind speed). Both these effects are common in complex terrain, and the SODAR makes boundary layer development easier to visualize, while as well giving a large volume of 3D numerical data.

Figures 8.12 and 8.13 show turbulent intensity (C_T^2) in complex terrain over a few hours. Figure 8.12 shows an overnight stable boundary layer situation with gravity waves in elevated layers. In Figure 8.13, the transition into a convective regime after sunrise is marked.

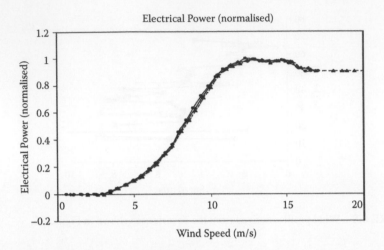

FIGURE 8.10 Power performance versus wind speed for mast-mounted cup anemometers (circular dots), SODAR (oblong dots), and ZephIR LIDAR (triangles).

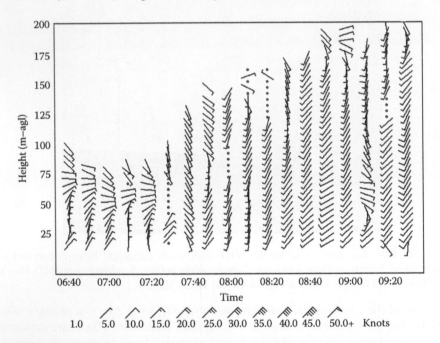

FIGURE 8.11 The velocity profiles observed by an AeroVironment 4000 SODAR in complex terrain.

8.1.5 Sound Speed Profiles

Outdoor sound propagation is increasingly important with noise sources such as airports, motorways, industry, and wind turbines increasingly being in close proximity to residential areas. In order to predict sound propagation over distances of

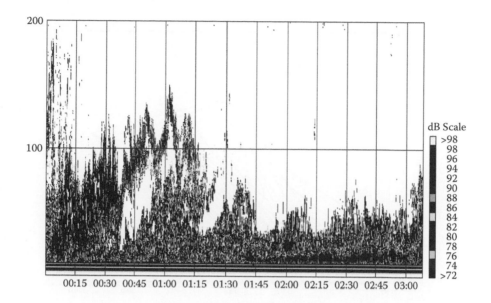

FIGURE 8.12 Turbulent intensity (C_T^2) during an overnight stable boundary layer situation. The vertical scale is height in m.

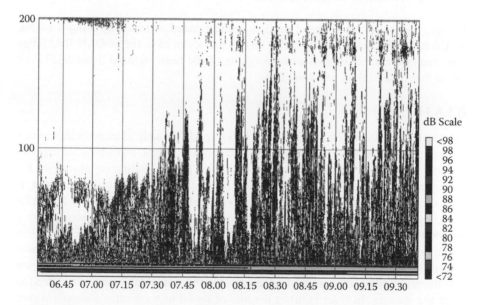

FIGURE 8.13 The transition from stable boundary layer to convective boundary layer. The vertical scale is height in m.

a few kilometers, it is necessary to know the atmospheric temperature and wind profile to perhaps 100 m. A SODAR/RASS combination can provide the necessary acoustic refractive index data on a continuous basis over a representative time scale.

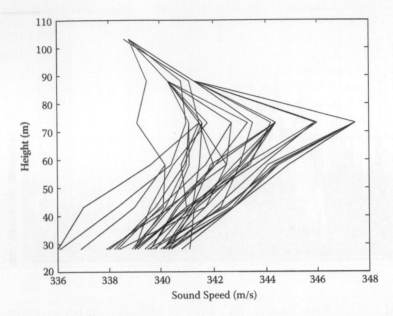

FIGURE 8.14   Sound speed profiles reconstructed from temperature and wind velocity profiles obtained from a Metek SODAR/RASS.

Figure 8.14 shows sound speed profiles reconstructed in this way using a Metek SODAR/RASS (Bradley et al., 2006). In this particular case, the SODAR/RASS has detected the presence of a jet which might not have been included in models based on surface observations and similarity.

8.1.6 HAZARDS

Increasingly SODARs and LIDARs are being used routinely at airports to monitor natural coherent wind structures (such as downbursts, gusts, and strong shear), and hazards caused by vortices from the wing tips of planes landing or taking off. By deploying an array of SODARs across the flight path, but outside the runway area, it is possible to obtain a "snapshot" of the entire wind field above the line of SODARs (Bradley et al., 2007). Figure 8.15 shows the vertical wind velocities recorded by a four-SODAR array during three aircraft landings. The SODARs were 25 m apart in a line on one side of the flight path. Spectral data were collected for single acoustic transmissions, every 2 s, rather than the normal averaging procedure. This meant that the acquired winds were not as accurate, but the fast update rate was required to track the vortices. In order to offset the loss of signal to noise ratio, a simple vortex model was fitted to the measured wind field every snapshot. This fitting of the velocity field was performed independently every 2 s, so smoothness of the estimated vortex movement and development was a strong indication that the method worked.

Figure 8.16 shows one example of the estimated development with time of the vortex-pair height and spacing, together with error bars. It can be seen that the method provides a good guide as to the vortex behavior.

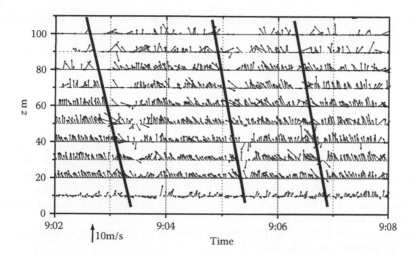

FIGURE 8.15 (a) Horizontal wind speeds measured at the four-beam SODAR. The orientation of the arrow indicates wind direction. A number of disturbances to the flow due to aircraft are shown by the solid sloping lines. (b) Typical vertical velocities measured by the four vertical-beam SODARs over a short period.

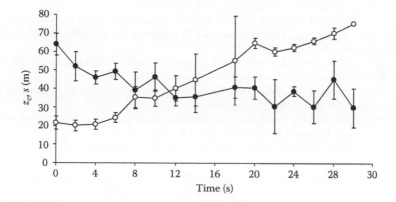

FIGURE 8.16 An example of the evolution of vortex height (filled circles) and half-spacing (open circles) in which the spacing increases substantially.

8.2   SUMMARY

In this chapter we have given a very brief coverage of some applications of SODAR and RASS. These indicate that

1. Acoustic remote sensing gives a very good visualisation of temporal development of wind and turbulence fields in the lowest few hundred meters.
2. Very good quantitative profiles and profile slopes are obtained even in difficult environments such as urban areas

3. Arrays of acoustic remote sensing instruments can give both vertical and horizontal temporal development, even on time scales of a few seconds
4. Highly accurate wind measurements are possible to support such endeavours as wind energy estimation

REFERENCES

Anderson PS (2003) Fine-scale structure observed in a stable atmospheric boundary layer by SODAR and kite-borne tethersonde. Boundary Layer Meteorol 107(2): 323–351.

Antoniou I, Jørgensen HE et al. (2004) Comparison of wind speed and power curve measurements using a cup anemometer, a LIDAR and a SODAR. EWEC-04, London.

Asimakopoulos DN (1994) Acoustic remote sensing and associated techniques of the atmosphere. Atmos Environ 28: 751–752.

Asimakopoulos DN, Helmis CG (1994) Recent advances on atmospheric acoustic sounding. Int J Remote Sens 15(2): 223–233.

Asimakopoulos DN, Helmis CG et al. (1996) Mini acoustic sounding – a powerful tool for ABL applications: recent advances and applications of acoustic mini-SODARS. Boundary Layer Meteorol 81(1): 49–61.

Barlow JF, Rooney GG et al. (2007) Relating urban boundary layer structure to upwind terrain for the Salfex campaign. Boundary Layer Meteorol.

Bradley SG, Antoniou I et al. (2004a) SODAR calibration for wind energy applications. Final reporting on WP3 EU WISE project NNE5-2001-297.

Bradley SG, von Hünerbein S et al. (2004b) High resolution wind speed profiles from a non-Doppler sodar array. 12th International Symposium on Acoustic Remote Sensing, Cambridge, UK.

Bradley SG, von Hunerbein S (2006) Use of arrays of acoustic radars to image atmospheric wind and turbulence. Inter-noise 2006, Honolulu, Hawaii, USA.

Bradley S, von Hünerbein S et al. (2006) Sound speed profile structure and variability measured over flat terrain. InterNoise, Hawaii.

Bradley SG, Mursch-Radlgruber E et al. (2007) Sodar measurements of wing vortex strength and position. J Atmos Ocean Technol 24: 141–155.

Coulter RL, Kallistratova MA (1999) The role of acoustic sounding in a high-technology era. Met Atmos Phys 71(1–2): 3–13.

Emeis S (2001) Vertical variation of frequency distributions of wind speed in and above the surface layer observed by Sodar. Meteorol Z 10(2): 141–149.

Engelbart D (1998) Determination of boundary layer parameters using wind-profiler/RASS and SODAR/RASS. 4th International Symposium on Tropospheric Profiling, Sowmass, Colorado.

Engelbart DAM, Steinhagen H (2001) Ground-based remote sensing of atmospheric parameters using integrated profiling stations. Phys Chem Earth Part B 26(3): 219–223.

Engelbart DAM, Steinhagen H et al. (1999) First results of measurements with a newly-designed phased-array Sodar with RASS. Met Atmos Phys 71(1–2): 61–68.

Helmis CG, Kalogiros JA et al. (2000) Estimation of potential-temperature gradient in turbulent stable layers using acoustic sounder measurements. Quart J Roy Meteor Soc 126(562A): 31–61.

Kirtzel HJ, Voelz E et al. (2000) RASS – a new remote sensing system for the surveillance of meteorological dispersion. Kerntechnik 65(4): 144–151.

Melas D, Abbate G et al. (2000) Estimation of meteorological parameters for air quality management: coupling of Sodar data with simple numerical models. J Appl Meteorol 39(4): 509–515.

Neisser J, Adam W et al. (2002) Atmospheric boundary layer monitoring at the Meteorological Observatory Lindenberg as a part of the "Lindenberg Column": facilities and selected results. Meteorol Z 11(4): 241–253.

Ostashev VE, Wilson DK (2000) Relative contributions from temperature and wind velocity fluctuations to the statistical moments of a sound field in a turbulent atmosphere. Acoustica 86(2): 260–268.

Peters G, Fischer B (2002) Parameterization of wind and turbulence profiles in the atmospheric boundary layer based on Sodar and sonic measurements. Meteorol Z 11(4): 255–266.

Piringer M, Baumann K (2001) Exploring the urban boundary layer by Sodar and tethersonde. Phys Chem Earth Part B 26(11–12): 881–885.

Raabe A, Arnold K et al. (2001) Near surface spatially averaged air temperature and wind speed determined by acoustic travel time tomography. Meteorol Z 10(1): 61–70.

Reitebuch O, Emeis S (1998) SODAR measurements for atmospheric research and environmental monitoring. Meteorol Z 7(1): 11–14.

Ruffieux D, Stübi R (2001) Wind profiler as a tool to check the ability of two NWP models to forecast winds above highly complex topography. Meteorol Z 10(6): 489–495.

Seibert P, Beyrich F et al. (2000) Review and intercomparison of operational methods for the determination of the mixing height. Atmos Environ 34(7): 1001–1027.

Singal SP (1997) Acoustic remote sensing applications. Springer-Verlag, New York.

Appendix 1

Mathematical Background

This book contains many equations, but in practice only very few mathematic concepts which are not straightforward algebra or calculus. In this appendix, we briefly review some of the frequently used signal-processing mathematics.

A1.1 COMPLEX EXPONENTIALS

Complex numbers are a compact method of describing vector quantities, which have both magnitude and direction. They can be visualized by considering an arrow pointing from 0 to 1 horizontally, or a unit vector $\underline{u}$. If distances from 0 to each position on this arrow are multiplied by −1, and the new positions plotted, the new arrow is simply a reversed version of the original (Fig. A1.1).

Multiplication of a vector by −1 is therefore equivalent to a rotation by 180°. Based on this concept, a rotation by 90° is implemented through multiplying by $\sqrt{-1}$, so that two successive multiplications by $\sqrt{-1}$ gives a rotation by 180°. Similarly, a rotation by 60° would be equivalent to multiplication by $(-1)^{1/3}$. Since $j = \sqrt{-1}$ the result is as shown in Figure A1.2.

A convenient way of describing a vector u which has both magnitude and direction is $u \cos\theta + ju \sin\theta$. This is called a complex number, with $u \cos\theta$ the real part and $ju \sin\theta$ the imaginary part.

In many cases we are interested in small changes in a vector $\underline{u}$, so need

$$\frac{d\underline{u}}{d\theta} = u\sin\theta + ju\cos\theta = -j\underline{u}$$

The *definition* of the exponential is that

$$\frac{de^{a\theta}}{d\theta} = ae^{a\theta}$$

so we can write

$$\underline{u} = ue^{j\theta}$$

The magnitude of the vector $\underline{u}$ is u and its argument is θ.

In the case of a wave varying sinusoidally with time, $\theta = \omega t$, and so the vector does a complete rotation in a time $2\pi/\omega$. In this context, $\theta = \omega t$ is called the phase.

FIGURE A1.1   Rotation of a vector by 180°.

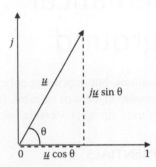

FIGURE A1.2   A vector in the complex plane.

A1.2   FOURIER TRANSFORMS

If a sine wave is multiplied by another sine wave of a different frequency, a composite wave is produced oscillating at the sum of the two original frequencies, but with its amplitude changing at a beat frequency equal to the difference of the original frequencies (Fig. A1.3). Also shown in Figure A1.3 is the mean value of the resulting waveform, averaged over the length of record shown. If the record is infinitely long, the mean value will be zero.

This is the technique used in mixing down or demodulating a Doppler-shifted signal to obtain a difference-frequency signal.

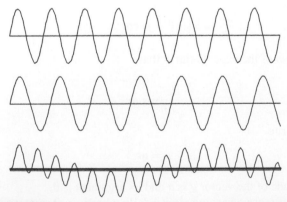

FIGURE A1.3   Multiplication of two sine waves to produce a beat frequency. The mean value over the length of record shown is the dark line in the lower plot.

If the two sine waves have the same frequency and phase as shown in Figure A1.4, the result of their multiplication is a sine wave at twice the frequency but everywhere positive. The mean value is then obviously positive, as shown.

Multiplication of a signal by a pure sine wave, and taking the mean of the result, tells us how close the pure sine wave is to the signal frequency. This is the principle of Fourier transforms.

However, the phase of the signal compared to the phase of the pure sine wave is also important. For example, Figure A1.5 shows multiplication of a sine wave by a sine wave of the same frequency but 180° out of phase. The mean value is just the negative of that in Figure A1.4.

Similarly, Figure A1.6 shows multiplication of two waves having the same frequency but a 90° phase difference. Now the mean value is zero.

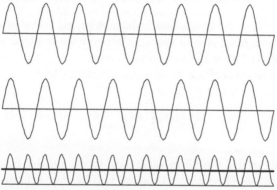

FIGURE A1.4 Multiplication of two identical sine waves produces a positive mean value (shown by the dark line in the lower plot).

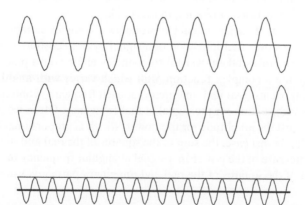

FIGURE A1.5 Multiplication of two sine waves of the same frequency but opposite phase. The zero line is shown in each plot, and the mean value of the product shown as a dark line in the lower plot.

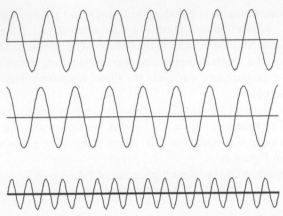

FIGURE A1.6   Multiplication of a sine wave and a cosine wave of the same frequency.

The phase variation can be allowed for by multiplying with $\cos(\omega t) + j\sin(\omega t)$ (or using kx if the signal is varying in space). Then both the in-phase and out-of-phase components are picked up in the averaging process.

The Fourier transform of a *general* signal $s(t)$ is therefore

$$S(\omega) = \int_{-}^{} s(t)e^{-j\omega t}\,dt$$

The averaging to find the mean value $S(\omega)$ for an angular frequency ω is performed by integration. Obviously, averaging will be over a finite time (or space) interval in practice. This gives the situation shown in Figure A1.3 where the mean value does not go to zero, even if the signal frequency is not the same as the pure sine wave frequency. The net result is that, even if a signal $s(t)$ contains a pure sine wave at angular frequency ω_0, the Fourier transform integrated over a finite portion of signal will respond with finite values $S(\omega)$ at frequencies near ω_0. This is the origin of the sinc function so often appearing in this book.

It is clear that the Fourier transform produces a complex number result, comprising the averages over multiplication of a signal by both $\cos(\omega t)$ and by $j\sin(\omega t)$. In general, the integral is taken over all frequencies ω (or over a practical range of frequencies), giving a complex function $S(\omega)$ which varies with angular frequency ω. The two components (real and imaginary) at each frequency contain both amplitude and phase information for the signal at that frequency. Often we are primarily concerned with just the amplitude (or the power, which is proportional to the square of the amplitude). In that case, the sum of the squares of the real and imaginary parts of $S(\omega)$ give a measure of the power in a signal at angular frequency ω, or the square root of the sum of the squares of the real and imaginary parts gives the amplitude.

A1.3   AUTOCORRELATION AND CONVOLUTION

It is clear form Figures A1.4–A1.6 that when two sine waves of identical frequency are multiplied, their relative phase determines the mean value of the result.

This gives a method for estimating when two signals are "lined up" and for estimating the time lag between them. So the cross-correlation between a signal $s(t)$ and a pure sine wave $\sin(\omega t + \varphi)$ of the same frequency is expressed as

$$\rho(\tau) = \int_{-\infty}^{\infty} s(t) \sin\left[\omega(t - \tau)\right] dt$$

where $\tau = \varphi/\omega$. More generally, the cross-correlation between signal $s(t)$ and another signal $q(t)$ is

$$\rho(\tau) = \int_{-\infty}^{\infty} s(t) q(t - \tau) dt$$

Because the signals are not generally pure sine waves, and the integral will be over a finite time span, $\rho(\tau)$ will vary over a range of τ values.

A useful special case is the autocorrelation, where

$$\rho(\tau) = \int_{-\infty}^{\infty} s(t) s(t - \tau) dt$$

This is a measure of how correlated one part of $s(t)$ is with another part separated by time τ. Spatial autocorrelations are also useful indicators of how quickly some spatially varying quantity is changing with distance.

A related integral is the convolution

$$c(\tau) = \int_{-\infty}^{\infty} s(t) q(\tau - t) dt$$

which arises when one signal is interacting with another but their relative phase is changing with time (such as when a transmitted signal moves over the spatially varying atmospheric reflectance profile).

Take the Fourier transform of $c(\tau)$:

$$C(\omega) = \int\limits_{-\infty}^{\infty} c(\tau) e^{-j\omega\tau} d\tau$$

$$= \int\limits_{-\infty}^{\infty} \left[\int\limits_{-\infty}^{\infty} s(t) q(\tau - t) dt \right] e^{-j\omega\tau} d\tau$$

$$= \int\limits_{-\infty}^{\infty} s(t) \left[\int\limits_{-\infty}^{\infty} q(\tau - t) e^{-j\omega\tau} d\tau \right] dt$$

$$= \int\limits_{-\infty}^{\infty} s(t) \left[\int\limits_{-\infty}^{\infty} q(\tau - t) e^{-j\omega(\tau-t)} d(\tau - t) \right] e^{-j\omega t} dt$$

$$= \int\limits_{-\infty}^{\infty} s(t) Q(\omega) e^{-j\omega t} dt$$

$$= S(\omega) Q(\omega)$$

This means that the Fourier transform of the convolution product of two signals is the product of the Fourier transform of one signal and the Fourier transform of the other signal. This is often useful.

A1.4   LEAST-SQUARES FITTING

Often we collect data points y_i with $i = 1, 2, \ldots, N$, corresponding to some changing condition, x_i. For example, y could be the wind speed estimated from a SODAR and x could be the wind speed measured by standard cup anemometers. The y values contain some variability due to random fluctuations, so it is useful to look for a simplifying model $y = f(x;a,b,\ldots)$, such as $y=ax+b$, which will summarize the results. It is important to note that the choice of the model is generally based on the assumption that the model describes the underlying physics in a reasonable way. So there might be, in some circumstances, a good reason to suspect a quadratic dependency between y and x, rather than a straight line dependency.

How can the unknown parameters a, b, ... be found? One common method is to minimize the average of the *squares* of the distances between the points y_i and the model prediction (ax_i+b for the straightline example). The *residuals* are

$$\varepsilon_i = y_i - f(x_i;a,b,\ldots)$$

and the sum of squares of residuals is

$$\chi^2 = \sum_{i=1}^{N} \varepsilon_i^2 = \sum_{i=1}^{N} \left[y_i - f\left(x_i; a, b, \ldots\right) \right]^2$$

The minimum of χ^2 is found by minimizing it with respect to each parameter, a, b, etc. as follows

$$\frac{\partial \chi^2}{\partial a} = \frac{\partial \chi^2}{\partial b} = \ldots = 0$$

For the model $y = ax$ (a straight line through the origin) this gives

$$\frac{\partial \chi^2}{\partial a} = \frac{\partial}{\partial a} \left[\sum_{i=1}^{N} y_i^2 - 2a \sum_{i=1}^{N} x_i y_i + a^2 \sum_{i=1}^{N} x_i^2 \right] = -2 \sum_{i=1}^{N} x_i y_i + 2a \sum_{i=1}^{N} x_i^2 = 0$$

$$a = \frac{\displaystyle\sum_{i=1}^{N} x_i y_i}{\displaystyle\sum_{i=1}^{N} x_i^2}$$

The sensitivity of this solution can be judged by seeing how the estimated value of slope a depends on variations in each y_i value. The result of all these dependencies gives the variance in a

$$\sigma_a^2 = \sum_{i=1}^{N} \left(\frac{\partial a}{\partial y_i} \right)^2 \sigma_{y_i}^2$$

A related measure of "goodness of fit" of the model to the data is the Pearson product moment correlation coefficient

$$r = \frac{\displaystyle\sum_{i=1}^{N} \left(x_i - \overline{x} \right)\left(y_i - \overline{y} \right)}{\sqrt{\displaystyle\sum_{i=1}^{N} \left(x_i - \overline{x} \right)^2 \sum_{i=1}^{N} \left(y_i - \overline{y} \right)^2}}$$

A matrix approach can also be taken so that sums like $\displaystyle\sum_{i=1}^{N} x_i y_i$ can be written in more compact form.

Appendix 2

Sample Data Sets and Matlab Code

The web page:

> http://www.phy.auckland.ac.nz/Staff/sgb/downloads

contains a number of Matlab m files (in the form of down-loadable txt files) which have been used to generate figures within the book chapters, together with two sample data sets: one (ASC_Data.txt) from an AeroVironment SODAR and the other (Metek_Data.txt) from a Metek SODAR/RASS.

Also included are Matlab m files, ASC_read.txt and Metek_read.txt to read each of the data sets and to produce plots of data. The manufacturers' data analysis and display software, which is much more comprehensive than the included Matlab files, are not provided since these are available to system purchasers under license.

It is hoped that these sample Matlab routines and data files will give the reader the opportunity to become familiar with the features of SODAR and RASS data, and to enable them to identify background noise and data quality issues.

Appendix 3

Available Systems

There are a number of prominent manufacturers of SODARs and RASS instruments. The following lists common systems available at the time of publishing. This is not an exhaustive list and is based solely on web page data.

NOTE: The specifications given below are those quoted by the manufacturer. Potential users of these systems are advised to also examine test data and independent intercomparisons, where available.

A3.1 AEROVIRONMENT INC. [CALIFORNIA, USA]

Although there are many AeroVironment systems in existence, the atmospheric remote-sensing sector of AeroVironment's business has been sold to a new company, Atmospheric Systems Corp. (ASC), described later.

A3.2 AQ SYSTEMS [STOCKHOLM, SWEDEN]

A3.2.1 AQ500 SODAR

The AQ500 SODAR comprises three independent parabolic dish segments, each with an individual speaker/microphone. A RASS system is also available.

Parameter	Value
Height	1.20 m
Width	1.00 m
Temperature range	−40 to +60°C
Humidity range	10–100% RH
Weight	38 kg
Antenna beam tilt	3 beams at 12°
Acoustic power (max)	4 W
Pulse power (max)	300 W
Transmitting frequency	2850–3550 Hz
Pulse repetition	Multimode
Altitude range	15–500 m
Height interval	5–25 m
Wind speed range	0–50 m/s horizontal ± 10 m/s vertical
Accuracy	0.1 m/s horizontal, 0.05 m/s vertical
Power requirement	12 VDC or 220 VAC
Power consumption	30–50 W

AQMR90		AQRASS	
Height	240 cm	Height	150 cm
Width	180 cm	Width	150 cm
Temperature range	−40 to +60°C	Temperature range	−40 to +60°C
Humidity range	10–100% RH	Humidity range	10–100% RH
Weight	150 kg	Weight	45 kg
Antenna beam tilt	0 and 15°	Focal length	710 mm
Acoustic power	<41 W	Power output	20 W
Pulse power (max)	300 W	Receiver type	Homodyne
Transmit frequency	1200–2800 Hz	Frequency	1290 MHz
Pulse repetition	Multimode	Polarization	Circular
Altitude range	25–1000 m	Altitude range	25–600 m
Height interval	10–50 m	Height interval	25 m
Wind speed range	Hor: 0–50, ver: ±10 m/s	Noise figure	<1.5 dB
Accuracy	Hor: 0.1, ver: 0.05 m/s	Accuracy	0.3 K T_v
Power requirement	12 VDC/220 VAC	Power requirement	220 VAC
Power consumption	150–200 W	Power consumption	120 W

A3.3 ATMOSPHERIC RESEARCH PTY. INC. [CANBERRA, AUSTRALIA]

Parameter	Performance
Horizontal wind components	Range 0–20 m/s, accuracy 0.2 m/s
Horizontal wind vectors	Range 0–25 m/s
Vertical wind components	Range 0–10 m/s, accuracy 0.1 m/s
Resolution of reading	0.1 m/s
Sampling height	5–290 m
Data interval	Approximately 1 min
Operating system	DOS or Windows
Display latency	Real time
Environmental conditions	−1 to +40°C, 0 to 100% humidity
Power supply	240 V
Acoustic frequency	4500–5000 Hz (selectable)
Dimensions (skid-mounted)	2400(l) × 1000(w) × 1200(h) mm

A RASS is also available, and larger low-frequency SODARs.

Parameter	RASS Performance
Temperature accuracy	0.5°C T_v
Height resolution	35 m
Height coverage	Greater than 700 m above ground level**
RF frequency	1270 MHz
RF power	15 W
Power supply	240 V, 50 Hz
Operating temperature	Field unit: −10 to +45°C
Operating temperature	Computer: +5 to +30°C
Relative humidity	0–90% non-condensing
Data interface	Serial interface, RS232
Data format	Virtual* temperature versus height
Data frequency	2 min, with running average on 10–15 min

A3.4   ATMOSPHERIC RESEARCH AND TECHNOLOGY LLC (ART) [HAWAII, USA] AND KAIJO CORPORATION [TOKYO, JAPAN]

See Figure A3.4. Also some specialized smaller versions are available.

A3.5   ATMOSPHERIC SYSTEMS CORPORATION (ASC) [CALIFORNIA, USA]

AeroVironment 4000/ASC SODAR	
Maximum altitude	200 m
Minimum altitude	15 m
Height resolution	5 m
Transmit frequency (approximate)	4500 Hz
Averaging interval	1–60 min (selectable)
Wind speed range	0–45 m/s
Wind speed accuracy	<0.5 m/s
Wind direction accuracy	±5°
Weight	116 kg
Antenna height	1.2 m
Antenna width	1.2 m
Antenna length	1.5 m

A3.6   METEK GMBH [ELMSHORN, GERMANY]

Mobile Doppler SODAR - MODOS

Wind velocity	0–35 m/s
Standard deviation of radial wind components	0–3 m/s
Finest height resolution	10 m
Minimum measuring height	30 m (10 m option)
Maximum measuring height (10 min averages)	Typical 200 m (90%); 500 m (60%)
Wind velocity accuracy	±0.5 m/s for 0–5 m/s
	±10% for 5–35 m/s
Wind direction accuracy	±5° for 0.8–35 m/s
Radial wind component accuracy	±0.1 m/s
Accuracy of standard deviations of radial wind	±0.15 m/s
Antenna	3 × 7 exponential horns
Aperture	1 m^2
Transmit power	1 kW (electric)
Beam tilt	Vertical; and 20°

	DSDPA.90-24	DSDPA.90-64
Frequency	1000–3000 Hz	1000–3000 Hz
Wind speed	0–50 m/s	0–35 m/s
Wind direction	0–360°	0–360°
Vertical wind speed	>±10 m/s	>±10 m/s
Operating temperature	30 to +55°C	–30°C to +55°C
Operating humidity	10–100 % (outdoor)	10–100 % (outdoor)
Integration time	Increments of 1 s	Increments of 1 s
Number of gates	adjustable, 1–50	adjustable, 1–50
Minimum measuring height	≥15 m, increment ≥1 m	≥15 m, increment ≥1 m
Height resolution	typically 10–30 m	Typically 10–30 m
Typical measuring height	350 m (70% availability)	500–800 m
Maximum measuring height	>1000 m	>1000 m
Recommended frequency	2200–2500 Hz	1500–2500 Hz
Signal power	Max. 800 W (elect.)	Max. 800 W (elect.)
Antenna gain	Typ. 20 dB	Typ. 25 dB
Sensitivity of receiver	10^{-6} N/m^2	10^{-6} N/m^2
Beam width	7–12°	Typ. 5–8°
Qualifying standards	DIN 3786 (11), KTA1508	DIN 3786 (11), KTA1508
Power consumption	250 W	250 W

A3.7   REMTECH SA [FRANCE]

	PA1	PA2	PA1-LR
Number of elements	52	196	52
Type of elements	Motorola 1025	Motorola 1025	Philips 3480/10
Nominal central operating frequency (Hertz) (five frequencies are emitted)	2100	2100	1050
Antenna size (m) (supporting structure not included)	0.65 × 0.65	1.3 × 1.3	1.4 × 1.4
Antenna weight (including supporting structure)	25 kg	100 kg	120 kg
Acoustic power	1 W	10 W	30 W
Maximum range	1000 m	1500 m	3500 m
Average range in typical conditions	550 m	1050 m	2500 m

For the RASS extension:

Radar Antennae (two at 5 m apart)

Diameter	2000 mm
Focal length	658 mm
Source type	Circular polarization
Height of the shielding fence around each antenna	2050 mm
Frequency	915 or 1290 MHz
Power	15 W
Bandwidth (at ?3 dB)	3 MHz
Rejection	Better than 60 dB
Total gain	39 dB
Noise figure	2 dB

A3.8 SCINTEC GMBH [TÜBINGEN, GERMANY]

	SFAS	MFAS	XFAS
No. of elements	64	64	52
Frequency (Hz)	2525–4850	1650-2750	825-1375
Electric(acoustic) power (W)	20 (2.5)	50(7.5)	500(35)
No. of frequencies	≤80	≤80	≤80
No. of beams	≤9	≤9	≤9
Beam angles	0°, ±19°, ±24°	0°, ±22°, ±29°	0°, ±22°, ±29°
Max. range gates	100	100	256
Vert. resolution (m)	5	10	20
Minimum range (m)	20	30	40
Maximum range (m)	500	1000	>2000
Averaging (min)	1–60	1–60	1–180
Accuracy horiz. wind (m/s)	0.1–0.3	0.1–0.3	0.1–0.3
Accuracy vert. wind (m/s)	0.03–0.1	0.03–0.1	0.03–0.1
Accuracy direction	2–3°		
Horiz. wind range (m/s)	0–50	0–50	0–50
Vert. wind range (m/s)	−10: +10	−10: +10	−10: +10
Operation temperature	−35: +50°C	−35: +50°C	−35: +50°C
DC power	±12 V, 4 A	±12 V, 8 A	±18 V, 20 A
AC power (W)	200	400	1500
Size (cm)	44 × 42 × 16	74 × 72 × 20	145 × 145 × 33
Weight (kg)	11.5	32	144

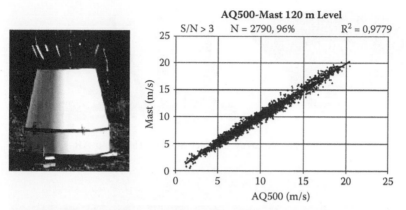

FIGURE A3.1   The AQ500 SODAR (left-hand image) and a comparison with mast anemometers (right-hand image).

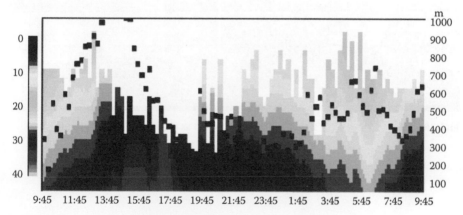

FIGURE A3.2   Typical AQ RASS output.

FIGURE A3.3   The ARPL mini-SODAR.

FIGURE A3.4 The KPA-1000 phased-array Doppler SODAR (left) and AR 410N series Doppler SODAR (right).

FIGURE A3.5 The ASC SODAR (left) and mounted as the ASC Wind Explorer (right).

FIGURE A3.6 Metek MODOS mobile SODAR.

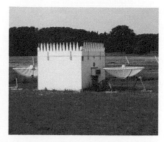

FIGURE A3.7 Phased-array SODAR DSDPA.90-24 (left) and phased-array SODAR DSDPA.90-64+RASS (right).

FIGURE A3.8   The REMTECH PA1 SODAR (left) and PA2 (right).

FIGURE A3.9   Scintec SODARs SFAS (left), MFAS (center), and XFAS (right).

Appendix 4

Acoustic Travel Time Tomography

There are infinitely many configurations possible, but for sample purposes we consider a square area surrounded on its perimeter by $4M$ evenly spaced sensor sites as in Figure A4.1. If transmissions along the boundaries are not included, each corner site transmits to $2(M-1)+1$ other sites. Each non-corner site transmits to $3(M-1)+2$ other sites. There are $P = 4[2(M-1)+1]+4(M-1)[3(M-1)+2] = 4M(3M-2)$ paths, or $2M(3M-2)$ bidirectional paths. Taking the example in Figure A4.1, $M = 2$ and there are 16 values of mean travel times from which to estimate temperatures. Consider a grid of 3×3 equal squares in which there are $N = 9$ temperature deviations from the base temperature T_0 (at which the speed of sound is c_0). If the square has sides of length L, the mean time of flight measurements can be written in the form

$$D = XT + E$$

where D is the $2M(3M-2) \times 1$ matrix of measured $c_0 \Delta t/L$ values, E is a matrix of measurement errors, T is the $N \times 1$ matrix of unknown $\Delta T/6T_0$ values, and X is

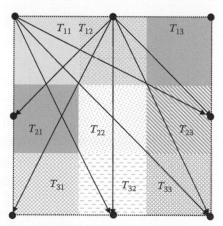

FIGURE A4.1 An example of eight sites and nine grid cells for travel time tomography.

$$
X = L
\begin{pmatrix}
\sqrt{5}/2 & 0 & 0 & \sqrt{5}/2 & 0 & 0 & 0 & \sqrt{5}/2 & 0 \\
\sqrt{2} & 0 & 0 & 0 & \sqrt{2} & 0 & 0 & 0 & \sqrt{2} \\
\sqrt{5}/2 & 0 & \sqrt{5}/2 & 0 & 0 & \sqrt{5}/2 & 0 & 0 & 0 \\
\sqrt{2}/2 & \sqrt{2}/2 & 0 & \sqrt{2}/2 & 0 & 0 & 0 & 0 & 0 \\
0 & \sqrt{5}/2 & 0 & \sqrt{5}/2 & 0 & 0 & \sqrt{5}/2 & 0 & 0 \\
0 & 1 & 0 & 0 & 1 & 0 & 0 & 1 & 0 \\
0 & \sqrt{5}/2 & 0 & 0 & 0 & \sqrt{5}/2 & 0 & 0 & \sqrt{5}/2 \\
0 & \sqrt{2}/2 & \sqrt{2}/2 & 0 & 0 & \sqrt{2}/2 & 0 & 0 & 0 \\
0 & \sqrt{5}/2 & \sqrt{5}/2 & \sqrt{5}/2 & 0 & 0 & 0 & 0 & 0 \\
0 & 0 & \sqrt{2} & 0 & \sqrt{2} & 0 & \sqrt{2} & 0 & 0 \\
0 & 0 & \sqrt{5}/2 & 0 & 0 & \sqrt{5}/2 & 0 & \sqrt{5}/2 & 0 \\
0 & 0 & 0 & 1 & 1 & 1 & 0 & 0 & 0 \\
0 & 0 & 0 & \sqrt{5}/2 & 0 & 0 & 0 & \sqrt{5}/2 & \sqrt{5}/2 \\
0 & 0 & 0 & \sqrt{2}/2 & 0 & 0 & \sqrt{2}/2 & \sqrt{2}/2 & 0 \\
0 & 0 & 0 & 0 & 0 & \sqrt{5}/2 & \sqrt{5}/2 & \sqrt{5}/2 & 0 \\
0 & 0 & 0 & 0 & 0 & \sqrt{2}/2 & 0 & \sqrt{2}/2 & \sqrt{2}/2
\end{pmatrix}
$$

The conventional least-squares solution is

$$
T = \left(X^T X \right)^{-1} X^T D - \left(X^T X \right)^{-1} X^T E
$$

This means the error in estimated $\Delta T/(6T_0)$ values are

$$
\frac{\sigma_T}{T_0} = 6 \frac{c_0 \sigma_{\Delta t}}{L} \left[\left(X^T X \right)^{-1} X^T I \right] = 6 \frac{c_0 \sigma_{\Delta t}}{L}
\begin{pmatrix}
0.2827 \\
0.4024 \\
0.2167 \\
0.3581 \\
0.2370 \\
0.4049 \\
0.2345 \\
0.3668 \\
0.2034
\end{pmatrix}
$$

In other words, the fractional error in temperature is comparable to $c_0 \sigma_{\Delta t}/L$. For example, in order to measure temperatures to 0.1°C within a 100 m × 100 m square, travel times need to be measured to within 0.1 × 100/(300 × 340) = 0.1 ms.

This example is only indicative that the challenge with acoustic travel time tomography is to measure the travel times to a small fraction of a millisecond and to do this simultaneously for many transmitting/receiving sites.

Appendix 5

Installation of a SODAR or RASS

SODARs and RASS instruments are designed to provide accurate measurements of boundary-layer parameters. But the specified accuracy is only possible if the instruments are set up correctly.

In this appendix, we include two documents:

1. A summary of the "Guidelines for the Use of SODAR in Wind Energy Resource Assessment" developed by Dr. Kathy Moore of Integrated Environmental Data, LLC and Bruce Bailey of AWS Truewind, LLC. This document has now been extended in consultation with a group attending the IEC meeting in Denmark, 2007.
2. Part of "Analysis of the AeroVironment MiniSODAR Data Processing Methods for Wind Energy Applications," a report on placement at the University of Salford as part of an MSc in Applied Physics by M. van Noord at the Delft University of Technology.

Although both documents are targeted at SODAR use for wind energy, they are indicative of the general considerations required for site selection and setting up an instrument.

A5.1 GUIDELINES FOR THE USE OF SODAR IN WIND ENERGY RESOURCE ASSESSMENT

A5.1.1 CALIBRATION AND TESTING

Since SODAR measures the wind speed in an elevated layer of air not typically accessed by other measurement systems (such as meteorological masts), calibration and test techniques often differ from those used for mechanical anemometry (EPA, 2000). The available techniques are described below:

1. SODAR *calibration and testing.* Some SODARs have calibration tools and techniques specific to the type or model of SODAR. In these cases, it is possible to test one or more of the following characteristics:
   - The SODAR array's response to known input frequencies. The results should be expressed as m/s wind speed per Hz of frequency shift. Check for both accuracy and resolution.

–   The output pulse length, frequency, and quality, to see if they conform
    to what they are supposed to be. Beam steering for phased-array sys-
    tems should also be confirmed by making phase angle measurements.
–   The condition and output of individual array elements (in phased-array
    systems) to ensure that all are operating properly.
–   To provide "challenge" input pulses with programmed delay and fre-
    quency (transponder test) corresponding to specific wind speed and
    direction at specific heights.
–   User-accessible test points where an oscilloscope can be used to check
    on the condition of electronic components.

2. *Comparison with mechanical anemometry on nearby tall masts.* SODARs,
   in general, must be placed at some distance from obstructions such as masts.
   When comparisons with tall towers are done as a means of calibration, the
   comparison should be done in simple terrain with low or at least uniform
   roughness. Additionally, the calibration of the mechanical anemometers
   must be well documented, and any sources of bias between the two resulting from differing measurement techniques, physics, and exposures must
   be accounted for (Bradley et al., 2005).
3. *Comparisons with rawinsonde data.* Comparisons of wind conditions measured by SODAR and rawinsondes are feasible, although balloon soundings
   of the atmosphere typically have low vertical resolution in the first 100 m
   above ground level. Balloons also move horizontally and vertically, and
   there will be low temporal resolution as well. Therefore this technique has
   limited application and is best done in areas consisting of simple, homogeneous terrain.
4. *Comparisons with tethered balloons.* Tethered-balloon systems equipped
   with a meteorological sensor package can also provide a general check on
   SODAR performance, although for wind energy resource assessment applications, this method is not sufficiently accurate for calibration purposes.

Calibration procedures and schedules should be documented thoroughly to support the use of SODAR in any wind resource assessment program. The documentation should include dates and locations of calibration tests, the names of personnel
involved, and the serial numbers of test equipment used (e.g., oscilloscopes or laptop
computers running test programs).

Calibration should be done at least every six months for any SODAR that is in
continuous use. The best practice is to calibrate every two to three months or before
any given measurement campaign, especially if long-distance transport or harsh
conditions have been experienced by the SODAR.

A5.1.2   OPERATING REQUIREMENTS

Retrieving and evaluating SODAR data daily using remote communications (digital,
analog, or satellite) is recommended. Some expertise and experience is required to
assess the quality of SODAR data.

SODAR should be operated at a site for a sufficient period of time to collect a representative and statistically robust sample of meteorological conditions for the desired range of wind speeds and directions. When comparing SODAR data with a reference wind measurement location, the data recording interval for both systems should be the same. Clocks within the data recorders for both systems should be synchronized.

Because the backscattered sound measured by SODAR is dependent upon spatially distributed turbulent temperature fluctuations, and these fluctuations are not necessarily evenly distributed within a height interval (i.e., range gate), very short measurement periods (less than a few days) are generally not very useful. Temporal averaging will smooth out the variation and provide better reliability and comparability with other measurements. Initial evaluation of the quality of SODAR data generally depends on at least 12 hours' data, preferably when wind speeds are 4 m/s or greater at the height level of interest (e.g., wind turbine hub height).

A5.1.2.1   Temperature

All SODARs require some kind of temperature setting or measurement as input. This setting allows the SODAR to accurately compute the speed of sound, which in turn determines both the altitude assigned to returned echoes, and, for phased-array systems, the vertical tilt of the acoustic beams. Because the SODAR determines the horizontal components from the component radial velocities in the tilted beams, the beam tilt angle variation with temperature can contribute to statistical error in the derived horizontal speed. Therefore, a realistic mean ambient temperature setting should be entered, or, if the temperature setting is updated automatically from a sensor logged with the SODAR, this option should be chosen in software.

A5.1.2.2   Precipitation

Precipitation can cause acoustic noise and/or scattering of sound back to the SODAR. For this reason, periods of precipitation should be removed from the SODAR data stream. In some SODARs, data acquisition can be automatically turned off when precipitation is sensed. For others, it is necessary to screen the data during post-processing in order to remove periods that are affected by rain or snow.

At mid-latitudes and high latitudes, a provision must be made for the removal of accumulated snow or ice from the SODAR's acoustic array and/or the reflector board. In some SODARs, a heater is provided which can be activated automatically when it snows. However, for SODARs operated off-grid, it may not be practical to provide sufficient power to do this. In this case, manual removal of snow will be necessary to maintain a quality data stream. Field notes should be kept on snow accumulation in the SODAR, so that data quality during those periods can be scrutinized. Even a light accumulation of snow can result in damped acoustic signals and poor altitude performance.

A5.1.2.3   Vertical Range and Resolution

The total possible vertical range of SODARs in common use for wind energy resource assessment varies from 200 to 500 m. The maximum possible height for a particular SODAR depends largely on the emitted power; however, the actual maximum height achieved at a particular site is determined by ambient atmospheric and noise conditions, and by the software settings, for example, the threshold for acceptable signal-to-noise ratio (SNR). Very dry or very noisy conditions, for example, will tend to limit the maximum achievable altitude performance.

SODAR produces acoustic pulses of discrete physical length (i.e., the pulse period in seconds times the speed of sound in m/s). The backscattered sound received from the atmosphere at any given time represents an integral of the sound through a depth related to the length of the pulse. The vertical resolution of the SODAR wind measurement, or the ability to distinguish between signals returned from different heights above the ground, depends primarily on three things: the acoustic pulse length, the sampling rate, and the number of samples required to convert from the time domain to the frequency domain (fast Fourier transform, or FFT). The choice of pulse length affects both the vertical resolution and the total height to which measurements can be made. The number of samples in each range gate affects the frequency resolution and hence the overall system accuracy.

SODAR users should be aware of the tradeoffs that are inherent in making choices between vertical resolution and frequency resolution. An optimal set of choices for any given instrument and measurement protocol should reflect this balance among altitude performance, frequency resolution, and vertical resolution.

Although most SODARs will output a data point for every 5 m depth, the actual vertical resolution is not better than 10 m (±5 m) in most circumstances, and it may be closer to 20 m (±10 m) because of the issues just cited. At adjacent range gates closer than the vertical resolution, there is "overlap" of information among the data. In a "regular" wind profile, the samples in the center of the range gate will tend to be weighted more than the samples at the extremes.

A5.1.2.4   Reliability Criteria

One output of most SODAR systems is a measure of the SNR, which is an indicator of data quality. In normal operation, SNR varies with the time of day because the amplitude (strength) of the backscattered acoustic pulse is dependent on the presence of turbulent temperature fluctuations. Periods when there is little or no sensible heat flux in the boundary layer (neutral conditions), and therefore little in the way of temperature variations, will produce less backscattered signal and lowered SNR. Low SNR can also result from very low humidity conditions. In addition, both acoustic noise and electronic noise can degrade the SNR or lead to false signals. Therefore, an important operating characteristic for SODAR is the SNR as one indicator of data quality.

Absolute values of the computed SNR vary with SODAR manufacturer, with site conditions, and atmospheric conditions. Plotting time series and vertical profiles of SNR can aid in establishing appropriate settings and the later identification of suspect data periods. The choice of threshold SNR to use for acceptable data depends to some degree on the site and conditions. A very noisy site may require a higher SNR to achieve quality data, while a quiet site may allow for a lower SNR threshold. When the SODAR is set up at a particular site, a good practice is to observe spectral data and SNR to determine if sufficient data of good quality are being acquired with acceptable altitude performance.

A5.1.3 Siting and Noise

SODAR should be located at a site that is representative of the prevailing wind conditions for the area of interest, similar to the way in which meteorological masts are sited. The SODAR should be placed on firm and level ground, and should be anchored if there is a risk of toppling due to high winds. SODAR siting must also take into account unwanted sources of ambient noise, fixed echoes, and sources of electrical noise, which can deteriorate data quality.

A5.1.3.1 Acoustic Noise (Passive and Active)

It is a good practice to develop an understanding of the acoustic environment in which the SODAR is operating, and optimize settings for that environment. When siting a SODAR system, consideration should be given to the location and spatial distribution of all potential acoustic sources and scatterers, whether atmospheric or not.

Fixed echoes, or passive noise, must be avoided when siting SODAR. Any back-scattered sound coming from fixed objects (masts, trees, buildings, etc.) is returned to the SODAR with zero Doppler shift. If this signal is as strong as or stronger than that from the atmosphere, the SODAR wind speed measurement will contain a low bias. Although most SODAR manufacturers provide software options for the detection and elimination of fixed echoes, the best practice is to avoid them in the first place by observing adequate setback distances, if at all possible. A starting point is to observe a setback distance no closer than the height of any fixed object in the vicinity. However, other considerations must be accounted for as well.

Knowledge of the SODAR's acoustic beam geometry can be used to orient the SODAR such that the side lobes strike objects such as buildings at an oblique angle. In this case, the echo is not reflected directly back to the SODAR, and the fixed echo effect can often be minimized. However, with obstacles such as trees, it can be difficult to find an orientation in which fixed echoes are not occurring. Trees also present a large amount of surface area for reflection of sound, which can result in multiple scattering of sound. The required distance to obstacles depends on the site, and how many obstacles there are (i.e., how much total surface area).

A further consideration is the SODAR beam geometry; a smaller tilt angle from the vertical for the horizontal velocity components should result in less interference from obstacles at lower heights. However, strong winds can refract (bend) acoustic beams, resulting in fixed echoes from objects which are theoretically below the main acoustic lobes. Some mitigation of the fixed echo effect may be achieved with additional acoustic baffling around the SODAR.

Detection of fixed echoes can be done by examining vertical profiles of wind speed, signal amplitude, and SNR. In general, the wind speed should increase with height, while the amplitude and SNR should decrease with height. A consistent deviation from these conditions at a particular height is diagnostic of a fixed echo.

Acoustic noise (active noise) can interfere with SODAR measurements by presenting false signals near the SODAR's acoustic frequency(ies) or by causing a degradation of the SNR, which results in degraded altitude performance. Active noise sources can include machinery such as generators or air conditioners, insects and birds, and even the wind itself blowing through and around trees or guy wires (Crescenti, 1998). Siting procedures should include an assessment of ambient acoustic noise using a noise meter or the SODAR in "listening" mode. Audio recordings, especially those done using the SODAR's antenna as the microphone input, can be very helpful in assessing the nature and impact of ambient noise. Acoustic shielding from some noise sources (e.g., generators) can be effectively obtained from bales of hay or other sound-absorbing material placed around the noise source.

A5.1.3.2   Electronic Noise

Input signals should be examined for the presence of radio frequency interference or other electronic noise produced by power supplies, inverters, communication equipment, fans, etc. If electronic noise is present, it can sometimes be diagnosed from audio files made through the SODAR antenna, or with the use of an oscilloscope at various test points in the SODAR. Depending on the source of electronic noise, it may be necessary to employ filters, shielding, or a different physical spacing of the electronic components in order to reduce it.

A5.1.3.3   Public Annoyance

The ongoing beeping or chirping of a SODAR can be an annoyance to people living nearby. If a SODAR is going to be operating 24 hours a day for some period of time, it is best to site it far enough from residences so as to minimize this annoyance.

A5.1.4   POWER SUPPLY AND SITE DOCUMENTATION

Although power consumption for most SODARs has decreased considerably in recent years, SODARs still consume more power than the typical mechanical anemometry used in wind resource assessment. Power should be sufficient to maintain continuous operation of the SODAR, as well as any communications equipment used for remote access of the instrument. If the SODAR is operated off-grid, some means of maintaining battery charge (generator, PV, wind generator) must be supplied. In mid-latitudes to high latitudes, PV charging that is sufficient in summer may have to be supplemented with another charging method in winter.

Site documentation should be similar to that done for meteorological mast measurements. Particular attention should be paid to the following items:

1. SODAR antenna rotation angle should be measured as accurately as possible (1°) with respect to true north. The SODAR should also be within 0.5° of level.
2. Any obstacles which could produce fixed echoes should be documented in an obstacle vista table with entries for azimuth, distance, elevation angle of the obstacle, and the degrees of arc occupied by the obstacle.
3. Any obstructions to wind flow and major changes in surface roughness should be noted for each wind direction.

4. The coordinates and elevation of the SODAR and co-located mast (if any) should be recorded.
5. The site's slope and aspect should be measured.
6. The distance and azimuth to any local mast used as a reference should be recorded.
7. Ambient noise sources should be noted and an audio record made, if possible.
8. There should be an onsite rain gage or precipitation sensor that is logged, to allow either for the suspension of SODAR measurements during precipitation, or the removal of those periods from the valid data set.

If a SODAR is used near existing wind turbine, it should be placed upwind in the prevailing direction.

A5.1.5   DATA COLLECTION AND PROCESSING

A SODAR's data outputs comprise the basic information sought by a SODAR measurement campaign to define certain atmospheric characteristics, such as wind shear. The robustness or temporal representativeness of the results depends on the duration of the measurement campaign and on the exclusion of precipitation periods. When comparing results of a SODAR campaign with results from another measurement system (e.g., mast), fundamental differences in measurement techniques must be accounted for.

A5.1.5.1   Data Parameters and Sampling/Recording Intervals

SODARs provide many output parameters. Primary outputs include all three component wind velocities (two horizontal and the vertical) and their standard deviations. In addition, some combination of the signal amplitude, noise amplitude, and/or the SNR, as well as the maximum height of reliable data, is also provided. There is a wealth of information both about the SODAR performance and about the meteorological conditions that can be derived from this assemblage of outputs. Some SODARs also provide estimates of meteorological parameters such as the standard deviation of the wind direction, the height of the inversion layer, the sensible heat flux, or the momentum flux. The degree of accuracy or reliability for derived parameters depends on the SODAR and the environmental conditions; it is best to check such derived parameters against other onsite instrumentation.

Recording intervals should be the same as those being used by other measurement systems with which the SODAR will be compared. Another consideration is that setting the maximum desired altitude affects the number of samples included in each recording interval. For example, setting the maximum altitude to 200 m results in about 15% fewer samples (chirps) per 10-minute recording interval, as compared to setting the maximum altitude to 150 m.

A5.1.5.2   Calculation of Wind Shear

SODAR produces a complete wind profile in the desired altitude range, with the actual height interval for data output determined by software settings, that is, every

10 m. As a result, the shear parameter can be determined between any two heights z_1 and z_2, and changes in the shear parameter can be detected throughout the measurement range. The shear parameter (α) is given by

$$\alpha = \frac{\log(U_2 / U_1)}{\log(z_2 / z_1)}$$

The underlying assumption in the use of the shear parameter for extrapolation is that a single power law wind profile pertains to the layer of interest. An alternative formulation of the wind profile, with more basis in physics, is the neutral logarithmic profile:

$$U(z) = \frac{u*}{\kappa_m} \ln(\frac{z}{z_0})$$

where $U(z)$ is the wind speed at any given height above the ground, u_* is the friction velocity, κ_m is the von Karman constant, and z_0 is a roughness length. Plotting SODAR wind speeds against the $\ln(z)$ can indicate

1. any non-uniformity in the wind profile which can be due to upwind changes in roughness or terrain,
2. the upper limit of the meteorological surface layer, that is, the layer for which the logarithmic profile pertains
3. the roughness length z_0 as a function of wind direction. Deviations from the logarithmic profile nearer to the surface can be accounted for with a displacement height parameter d:

$$U(z) = \frac{u*}{\kappa_m} \ln(\frac{z-d}{z_0})$$

This parameterization, mainly used as a convenience for computational purposes, is commonly interpreted to mean that the displacement height is a kind of "virtual surface" representing the mean height of momentum absorption.

As with shear calculations based on mechanical anemometry, only wind speeds greater than 4 m s^{-1} should be considered in the calculation of the wind shear at a site for wind energy purposes. This is the wind speed below which most wind turbines do not produce power.

A5.1.5.3   Measurement Period

The decision about how much SODAR data to collect at a site depends on the objective of the study. For example, one criterion might be to collect enough qualified data to achieve 95% confidence bounds of ±0.02 around the mean shear exponent for the prevailing wind direction sector(s). This criterion can often be achieved in as little as three weeks, if there are enough observations with wind of sufficient speed from the important energy-producing wind direction sectors. However, in other cases, either a longer period might be required or measurement periods in different seasons may to necessary to achieve sufficiently representative data.

Another approach to determining the duration of a SODAR study is to achieve target confidence limits around the speed difference with a reference meteorological mast located some distance from the SODAR site. Whether the criterion is based on confidence bounds around the shear or the speed differences, it is best to use whatever pre-existing information there is about a site, such as seasonal variability in shear, atmospheric stability, or wind direction distribution, to help determine the period duration (and number of seasonal periods) needed to achieve a representative data set.

A5.1.5.4   Exclusion of Precipitation Periods

Acoustic signals can be scattered back to the SODAR from hydrometeors (rain drops or snow flakes), depending on the acoustic frequency. In addition, there can be noise from raindrops striking the exposed area of antenna or transceiver. It is best to check with the manufacturer regarding the effect of precipitation on SODAR data quality; for most SODARs, periods of precipitation must be removed from the data. Even after removal of periods with recorded rain or snowfall, the data should be screened for periods of excessive negative vertical velocity, which may indicate that light precipitation, unrecorded by a gauge, was occurring. Such screening can be achieved through examination of the time series of vertical velocity.

A5.1.6   COMPARISONS WITH MECHANICAL ANEMOMETRY

Since SODAR data is almost always referenced to ongoing mechanical anemometry, whether co-located or at some distance from the SODAR site, bias between mechanical anemometry and SODAR wind speeds arising from differing underlying physics between the two should be addressed in data processing. Bias between the two can be attributed to several factors:

- SODARs generally report a vector-average wind speed, while cup anemometers yield a scalar average. The vector average can be as much as 5% lower than the scalar average, but the median difference is generally closer to 2% to 3%. A conversion between the two can be made using the standard deviation of the wind direction, if a mechanical wind vane is present, or using an empirical relationship with the SODAR sigma-w (standard deviation of the vertical wind velocity), for instance.
- The tilt angle of the off-vertical acoustic beams of the SODAR phased array is subject to variation due to temperature. Most SODARs have an ambient temperature setting which is generally set to some average value for the period of measurement. However, depending on the acoustic beam geometry, greater wind speed measurement accuracy will be achieved if *either* a correction based on the temperature–tilt angle relationship is made after the data are collected, *or* an in situ temperature measurement is used to calculate and adjust the tilt angle as the data are being collected.
- Mechanical anemometers can overestimate the wind speed due to overspeeding resulting from turbulence or off-horizontal flow. The magnitude and char-

acteristics of overspeeding vary with the anemometer. These effects should be accounted for when making comparisons between SODAR and anemometers.

- The geometry of the SODAR acoustic beams may be such that in cases where there is an inhomogeneous wind field, the different off-vertical beams may be probing flows of different characteristics. This could be detected by gross discrepancies in the wind direction, compared to a nearby tower.
- SODAR calculates the wind speed in a volume of air, in contrast to the "point" measurement provided by mechanical anemometers. If the wind profile has very high shear in it, this will cause the SODAR speed at the lowest heights to be lower than a point measurement centered in the SODAR volume, by as much as 3% to 4%. This will result in a concomitant increase in the shear of as much as 5%, depending on the surface roughness. If the shear decreases with increasing height, then this effect will also decrease with height. This example supposes a volume average of 20 m in depth, and a 50/30 m shear parameter (point measurement) of 0.40.

Comparisons between SODAR and mechanical anemometry should include a careful examination and verification of the location and characteristics of the ane-mometry. Characteristics to examine or verify include measurement levels of mast-mounted sensors, directional orientations of sensor booms from the mast, distances between sensors and mast hardware, sensor calibration constants, changes in instru-mentation, etc. Valid measurements for the anemometry should exclude periods of detectable mast-induced flow interference (e.g., tower shadow), periods when icing is occurring, or periods when other types of measurement problems are occurring.

A5.1.7   Other Considerations for Incorporating SODAR Information into a Resource Assessment Program

For many applications, SODAR will likely be used for relatively short periods of time and the results compared to longer-period measurements taken by ongoing meteoro-logical masts. Evaluation of the seasonal representativeness of the SODAR measure-ment period should be done by examining the seasonal changes in the shear at the site of interest. In many cases no adjustment need be made, but in cases where the shear is expected to change significantly by season, SODAR should be deployed accordingly.

Beyond the factors described above, significant discrepancies between results obtained by SODAR and conventional mast anemometry may still occur and under-standing of the source(s) of such discrepancies should be sought. An obvious potential source of discrepancy can be the separation distance between the two measurement systems and the corresponding differences in upwind fetch, surrounding surface roughness, ground-base elevation, and other physical factors. Differences in loca-tion between measurement systems, therefore, must be accounted for when utilizing SODAR to estimate wind shear conditions above existing meteorological masts.

TABLE A5. 1

SODAR parameters of interest

Parameter	Description
FREQ	Sound transmit frequency (f_T)
AMP	If BACK is equal to 0, then AMP is the minimum acceptable signal amplitude in mV, otherwise the feature is disabled
BACK	Percentage of the background noise sample (in mV) used to threshold the acceptable signal amplitudes; if BACK is greater than 0, then the noise is sampled prior to the pulse; if BACK is less than 0, the noise is sampled at the end of the pulse; if BACK is equal to zero, AMP becomes the threshold used to qualify good samples
BW	The analogue fourth-order filter bandwidth in Hz
CLUT-	The beams for which the clutter rejection algorithm is activated; the vertical beam is 1, the y beam is 2, and the x beam is 4; beam combinations are identified by summing the beam designators
PULW	Pulse length in milliseconds (τ)
RISE-	The pulse rise and fall time in the number of SRATE samples
MHT	Maximum sampling altitude (default = 200 m)
SEC	Averaging time for wind value calculation (default = 150 s)
GD	Percentage of good data samples for each beam and each altitude needed to produce a valid Doppler shift estimate (default = 5%)
WMAX	The maximum acceptable vertical velocity estimate (in cm/s) used to correct the horizontal wind component data; if the vertical velocities exceed this value, then the ASP assumes that the vertical wind correction is 0 cm/s; if WMX is 0, then no correction is made for the vertical velocity (default = 500 cm/s)
SRATE	The Doppler channel analogue-to-digital conversion rate or sampling rate in Hz (f_s) (default = 960 Hz)
NFFT	The number of data points used for spectral processing at each range gate (N_s) (default = 64)
DAMP	Percent of maximum pulse amplitude used for operation (default = 100)
TILTC, TILTB	The tilt angles (θ) for the B and C (2 and 3) beams (antennas); these angles are dependent on FREQ (default = 16°)
MINAR, MINBR, MINCR	The minimum radial velocity accepted for the axes 3, 2, 1, respectively (default = −400, −800, −800 cm/s)
MAXAR, MAXBR, MAXCR	The maximum radial velocity accepted for the axes 3, 2, 1, respectively (default = 400, 800, 800 cm/s)
SPECS	Selects the first range gate (0, 1, ...) for which the spectral moments and the spectral data will be recorded (default = 2)
SPECI	Selects the spacing between the altitudes recorded; 1 = every gate, 2 = every second gate(default = 1)
SPECL	Number of range gates (default = 40)

A5.2   ANALYSIS OF THE AEROVIRONMENT 4000
SODAR DATA PROCESSING METHODS

A5.2.1   PARAMETER SETTINGS

The settings of the SODAR parameters are given in Table A5.1.

A5.2.2   INTERRELATIONS OF PARAMETERS AND
CONDITIONS FOR PARAMETER VALUES

Although the SODAR allows for a lot of different parameter settings, a number of these parameters are interrelated. In some cases, this means that it is not possible to set one parameter to one value and another parameter to a particular other value. In other cases, it just means that it is very unpractical or unbeneficial to do this. Which parameters are interrelated, and how they are, is described below.

MHT, SEC, and GD: The MHT defines the pulse repetition rate of the SODAR, and SEC is the averaging time for the calculation of a single wind speed. These two parameters together define the number of points averaged in each calculation. With MHT = 200 m, the pulse repetition time is just below four seconds. This means that about 38 data points are averaged to find one wind speed value. If the percentage of good data samples GD equals 5%, the criterion for the computation is that at least two data points should be used. This implies a huge relative uncertainty of 0.7. For a relative uncertainty of 50%, which is still large, GD would have to be set to 10% if MHT and SEC remain the same.

FREQ and TILTB/C: As the tilted beams are created with a phased array, their tilt angle depends on the transmit frequency. For the AeroVironment SODAR, transmitting at 4500 Hz, this means that the tilt angle will be 16° at all times.

RISE, SRATE and PULW: By setting RISE to a finite number, the amplitude of the sound pulse will be Hanning-shaped. The benefit of doing this is that two pulses can be distinguished even when they overlap partially. The pulse amplitude has to have fallen off to half of its maximum value before the new pulse starts in order to distinguish the two. Effectively this means that τ decreases from PULW into PULW $- (2 / 3)$(RISE/SRATE). Note that RISE should always be equal to or smaller than SRATE/2. If they are equal, the pulse will be a full Hanning shape.

PULW, SRATE and NFFT: As was discussed in Section 2.1.3 the height and the wind speed resolution are both minimized when $\tau = N_s / f_s$ is chosen. In SODAR parameters, this means: PULW = NFFT/SRATE. The exact value of τ to be chosen depends on the requirements for the results. High τ will give a good spectral resolution, but low spatial resolution, whereas low τ will have the opposite effect.

MINR, MAXR, FREQ, SRATE and BW: The limits for the accepted radial wind speeds should fall within the measurable range of speeds. Theoretically, the measurable speed is limited by the maximum Doppler shift, which equals $\pm f_s/2 = \pm$SRATE/2. In practice, a peak at this frequency would result in aliasing, and therefore the spectrum is cut off at a frequency just smaller than $\pm f_s/2$, by setting BW just smaller than SRATE. The maximum measurable Doppler shift is then $\pm$BW/2, so the maximum measurable radial wind speed is $V_{r,\max} = \mp \dfrac{BW \cdot c}{4 f_T}$. Then it is required that $MINR \geq -\dfrac{BW \cdot c}{4 f_T}$ and $MAXR \leq +\dfrac{BW \cdot c}{4 f_T}$.

At the same time, it is important to realize what horizontal and/or vertical winds one wants to be able to measure. With the parameter values as listed in Table A5.1, one can measure a maximum vertical wind speed of 4 m/s and a maximum horizontal wind speed (absolute values) between 15 and 29 m/s for a vertical wind of 0 and 4 m/s, respectively.

MHT: The maximum sampling altitude has to be set to a value significantly higher than the maximum height one actually wants (or expects) to receive data from. The reason for this is that MHT defines the pulse repetition time of the SODAR and if this time is set too small, data from high-range gates can be received during the listening mode following the *next* pulse. The SODAR will then interpret this as a signal from a low-range gate (according to the time between the "next" pulse and the receiving of the signal) and might return a wrong value of the wind speed.

Furthermore after sampling the range gate at MHT, the SODAR continues to sample the received signal for a short while. It assumes that at this point there is no weather signal, so that the signal consists of background noise only, and therefore uses this data for estimating the background noise. Now, in case there is a weather signal, this will be assumed to be noise and increase the background noise level. This might then cause several good spectra to be rejected since they do not meet the SNR requirements.

WMAX and MAXCR, MINCR: As MAXCR and MINCR limit the acceptable vertical wind speed for each individual spectrum, setting WMAX to a higher absolute value than the first two parameters does not introduce any new limitations. Setting MAX/MINCR to the same absolute value as WMAX might give different results than setting them higher, as this influences the results of the averaged wind speeds and WMAX only works on the latter.

A5.2.3 Use of an Artificial Signal to Verify Performance

In this section we describe methods by which some elements of SODAR operation can be checked. A speaker is connected to the output of a wave generator and is situated at a position at the top of the SODAR baffle. The sound emitted by the speaker is then recorded and processed by the SODAR. At the time of the measurements, the SODAR is not emitting any sound itself, but is only "listening." The first part of the data processing to inspect is the FFT of the SODAR time series. In order to check its applicability and its dependency on the SODAR parameters, several acoustic signals are recorded and transformed by the SODAR. These signals are produced by the wave generator. Each signal is analyzed by the SODAR for different parameter settings or speaker positions. Each measurement is taken for 15 minutes and then averaged over this time, to get a statistically acceptable spectrum. With the SODAR parameter settings used, this gives to an average over 230 spectra. So the relative uncertainty in any of the spectral densities is

$$\frac{1}{\sqrt{230}} = 0.07$$

The first two measurements (Tables A5.2 and A5.3) focus on the SODAR parameters that play a role in the discrimination of peak and noise. As the input signal in this case is white noise, the expected signal is a flat, uniform one. By averaging the calculated spectra over 15 minutes and comparing them for different parameter settings, the influence of these settings on the spectra is found.

TABLE A5.2
Measurement 1: signal amplitude settings

ID	Input signal	AMP	BACK
1.1	White noise (V_{pk}=19 mV)	15	120
1.2	White noise (V_{pk}=19 mV)	15	−120
1.3	White noise (V_{pk}=19 mV)	15	−170
1.4	White noise (V_{pk}=19 mV)	100	0
1.5	White noise (V_{pk}=19 mV)	300	0

TABLE A5.3
Measurement 2: signal to noise and clutter rejection settings

ID	Input signal	SNR	CLUT
2.1	White noise (V_{pk}=15 mV)	0	0
2.2	White noise (V_{pk}=15 mV)	1	0
2.3	White noise (V_{pk}=15 mV)	5	0
2.4	White noise (V_{pk}=15 mV)	1	1
2.5	White noise (V_{pk}=15 mV)	1	2
2.6	White noise (V_{pk}=15 mV)	1	3
2.7	White noise (V_{pk}=15 mV)	1	4
2.8	White noise (V_{pk}=15 mV)	1	5
2.9	White noise (V_{pk}=15 mV)	1	6
2.10	White noise (V_{pk}=15 mV)	1	7

In the third measurement (Table A5.4), the effect of the analogue bandwidth settings is investigated. Is this specific bandwidth the only one that influences the spectrum or is there another fixed bandwidth that influences it?

TABLE A5.4

Measurement 3: analogue bandwidth settings

ID	Input signal	BW (Hz)
3.1	White noise (V_{pk}=15 mV)	1000
3.2	White noise (V_{pk}=15 mV)	800
3.3	White noise (V_{pk}=15 mV)	600

TABLE A5.5

Measurement 4: FFT center frequency

ID	Input signal	FREQ (Hz)
4.1	White noise (V_{pk}=15 mV)	4700
4.2	White noise (V_{pk}=15 mV)	4900
4.3	White noise (V_{pk}=15 mV)	4500

For measurement 4, the SODAR setting for the transmit frequency is changed (Table A5.5). Although the SODAR is not actually emitting any sound, this setting changes the center frequency of the spectra, while the bandwidth is not changed. As the input signal is white noise, the spectra should all be flat, no matter what the frequency setting is. If they are not flat, but for example show a broad peak at equal distances from the center frequency, this would indicate a systematic error in the calculation.

The final two measurements, 5 and 6, do not change the SODAR parameters, but the input signal itself changes. In the first case (Table A5.6), the input is a pure sine wave, hence it should result in a single spectrum peak. Any significant distinction from that would show an error in the spectrum calculation. The second (Table A5.7) does not change the signal generated by the wave generator, but it changes the position of the signal speaker. This is done to check the influence of the speaker position on the spectrum and to take that into account for all other measurements.

TABLE A5.6

Measurement 5: sine wave input

ID	Input signal	Frequency (Hz)
5.1	Sine wave (V_{pk}=3.05 mV)	4100
5.2	Sine wave (V_{pk}=3.05 mV)	4350
5.3	Sine wave (V_{pk}=3.05 mV)	4600
5.4	Sine wave (V_{pk}=3.05 mV)	4850

TABLE A5.7

Measurement 6: speaker position

ID	Input signal	Speaker position
6.1	White noise (V_{pk}=15 mV)	Opposite reflector
6.2	White noise (V_{pk}=15 mV)	On one side
6.3	White noise (V_{pk}=15 mV)	Top of reflector

The first step is to check the peak detection using a single spectral peak at a known frequency with minimum noise. Therefore, the spectra of measurement ID's 5.2 and 5.3 (Table A5.6) are used. For the noise level, a similar approach is used as for the peak frequency. The simple case now is white noise, as the assumption in all of the algorithms is that the noise is white noise. The spectrum of measurement 6, position 2, is used for this case. The noise level as estimated by the SODAR is not one of the output parameters, but can be derived from two other parameters: the peak amplitude and the signal to noise ratio for each spectrum. Many plots of spectra are generated in this process. The results of one set of such measurements on one particular instrument are summarized in Table A5.8.

TABLE A5.8

Conclusions drawn on the results of the six artificial input measurements

Measurement	Conclusion
1. Signal amplitude	The settings of AMP and BACK have no effect on the resulting spectrum
2. SNR and clutter rejection	SNR and CLUT settings do not influence the resulting spectrum
3. Analogue bandwidth	BW defines the dominating filtering bandwidth for the spectrum; this setting should be used to prevent aliasing; an 800 Hz bandwidth seems to suffice
4. FFT center frequency	The FFT center frequency, defined by FREQ, does not influence the spectra
5. Sine wave input	Input sine waves with a fixed amplitude can result in spectral peaks with different amplitudes, depending on their location in the spectrum; this might result in the rejection of otherwise acceptable spectra
6. Speaker position	The speaker position influences the spectrum due to phase differences; however, for normal operation effects like this only show for fixed echoes; for returned wind signals the wave front is practically flat and the SODAR corrects for the beam direction; in future measurements with speakers, the speakers should be placed in the far field of the antenna

Overall, one can conclude that only the setting for the analogue bandwidth influences the computation of the spectra by the SODAR. This influence is expected and setting this parameter to the right value is necessary to create useful spectra.

REFERENCES

Antoniou I, Jorgensen HE, Ormell F, Bradley SG, vonHunerbein S, Emeis S, and Warmbier G (2003) On the theory of SODAR measurement techniques. RISO-R-1410 (EN), 59 pp.

Bradley S, Antoniou I, vonHunerbein S, Kindler D, deNoord M, Jorgensen H (2005) SODAR calibration for wind energy applications. Final reporting on WP3, EU WISE project NNES-2001-297. The University of Salford, Greater Manchester, UK, March 2005, 69 pp.

Coulter RL, Kallistratova MA (1999) The role of acoustic sounding in a high-technology era. Meteorol. Atmos. Phys. 71: 3–13.

Crescenti GH (1998) The degradation of Doppler SODAR performance due to noise: a review. Atmos. Environ. 32(9): 1499–1509.

Crescenti GH (1997) A look back on two decades of Doppler sodar comparison studies. Bull Am Met Soc. 78(4): 651–673.

EPA (2000) Meteorological Monitoring Guidance for Regulatory Modeling Applications. EPA-454/R-99-005. U.S. Environmental Protection Agency, Research Triangle Park, NC.

Index